本译丛受到宁波大学科学技术学院重点学科（翻译专业学位硕士点）培育项目资助，为译行丝路·翻译研究与语言服务中心系列成果之一

**Translating Ningbo**

Editors   Xia XIANG   Anfang HE

# Architectural Wonders in Ningbo Land

## 奇构巧筑　宁波建筑文化

Author

Dingfu HUANG

Translators

Zhihui MAO

Rui XU

Reviewer

Tiezhu DONG

ZHEJIANG UNIVERSITY PRESS
浙江大学出版社
·杭州·

宁波出版社
NINGBO PUBLISHING HOUSE
·宁波·

**图书在版编目（CIP）数据**

奇构巧筑：宁波建筑文化 = Architectural
Wonders in Ningbo Land：英文 / 黄定福著；毛智慧，
徐睿译. -- 杭州：浙江大学出版社，2025. 4.
（宁波文化译丛 / 项霞，贺安芳总主编）. -- ISBN 978
-7-308-25586-8

Ⅰ. TU-092

中国国家版本馆 CIP 数据核字第 20249WS068 号

奇构巧筑：宁波建筑文化
Architectural Wonders in Ningbo Land

黄定福　著　毛智慧　徐　睿　译　董铁柱　译审

策划编辑　黄静芬
责任编辑　黄静芬
责任校对　杨诗怡
封面设计　周　灵
出版发行　浙江大学出版社
　　　　　（杭州市天目山路148号　邮政编码310007）
　　　　　（网址：http://www.zjupress.com）
排　　版　杭州林智广告有限公司
印　　刷　浙江新华数码印务有限公司
开　　本　710mm×1000mm　1/16
印　　张　17
插　　页　4
字　　数　345千
版 印 次　2025年4月第1版　2025年4月第1次印刷
书　　号　ISBN 978-7-308-25586-8
定　　价　78.00元

Stilt Architecture in the Hemudu Site

Xiantong Pagoda

Great Hall of Baoguo Temple

Yingzhou Jiewu Arch

Bailiang Bridge

The Site of Yongfeng Warehouse

Feng Yue's Painted *Taimen*

Yongshang Zhengren Academy

Shuangzhimiao Stage

Danai Hall

Qing'an Guild Hall

The Drum Tower

Lingxing Gate and Panchi Bridge in the Cicheng Confucius Temple

The Former Site of Zhejiang Customs

*Shikumen* Architecture

Tianxu Hall in Yu Qiaqing's Former Residence

# 总　序

宁波历史悠久，文化璀璨；宁波历史文化研究著述丰富，成绩斐然。然而，如何让宁波文化走出国门，如何向世界讲好宁波故事，却一直是个薄弱环节。这与宁波自古以来就崇尚开放的城市禀性不够合拍，也与今天"滨海宁波，扬帆世界"的时代追求不相协调。

有鉴于此，宁波大学科学技术学院人文学院翻译专业教师团队大胆提出：将宁波出版社2014年精心推出的"宁波文化丛书"（第一辑）翻译成英文版，让这套丛书作为文化使者，把宁波历史、宁波文明、宁波精神、宁波智慧带到四面八方。

"宁波文化丛书"（第一辑）共8本，出于一些原因，我们替换了其中1本，因此这套"宁波文化译丛"也是8本。具体是：王耀成著《商行四海：解读宁波帮》，黄定福著《奇构巧筑：宁波建筑文化》，谢安良著《丝路听潮：海上丝绸之路文化》，黄渭金著《东方曙光：宁波史前文明》，涂师平著《羽人竞渡：宁波发展史话》，方同义著《千年文脉：浙东学术文化》，虞浩旭、张爱妮著《甬藏书香：宁波藏书文化》，郁伟年著《阿拉宁波人》。

得益于原作者在挖掘、整理和研究宁波历史文化方面所做出的艰苦努力，文质兼美的文本为我们的翻译工作奠定了良好的基础。

但是，要忠实地翻译原著并不是一件容易的事情。以古文翻译为例，上述著作都或多或少征引了古代文献，有的不仅数量很多，而且专业性很强。对于古代文献，中文著作往往直接引用原文，至于古文如何解读可以让读者自己去琢磨。译著则不同，必须做到字字落实，不能含糊，所以翻译起来特别费时费力。我们还经常碰到引用的古文有文字错讹的情况，那就更加考验译者的学问和识见了。一边做翻译，一边查资料，一本书翻译下来，硬是把英语老师磨炼成古文通了。

这只是翻译三要素"信、达、雅"中"信"的问题。而译者追求的不仅仅是"信"，还有"达"与"雅"。翻译不是简单的语言转换，而是一种再创造。为了向读者奉献一套质量上乘的译丛，我们采取译者精心打磨、同事互相切磋、专家审读把关等措施，把精品意识贯穿于翻译工作的全过程。我们希望我们的译作对得起优秀的宁波文化，也对得起自己的职业操守。

在译丛即将面世之际，我们衷心感谢原作者的创造性劳动，感谢浙江大学出版社和宁波出版社的大力支持，感谢审稿专家和责任编辑的辛勤付出，感谢项霞、贺安芳两位总主编的精心组织和策划。

限于时间和水平，这套译丛也可能存在这样那样的不足甚至错误，欢迎专家和读者提出批评意见。

本译丛受到宁波大学科学技术学院重点学科（翻译专业学位硕士点）培育项目的资助，由译行丝路·翻译研究与语言服务中心出品，是宁波大学科学技术学院翻译团队"翻译宁波"（"Translating Ningbo"）的第一步。万事开头难，有了第一步就可以有第二步、第三步……我们将继续努力，为传播宁波历史文化做出自己应有的贡献。

周志锋

2022 年 11 月 11 日

于宁波大学科学技术学院

# CONTENTS

# A Review of Ningbo Architectural Culture:
# Creativity and Harmony in Construction

Architecture is the mirror of history.

Ningbo, a famous historical and cultural city, has a long history and a long-standing culture. Located at the estuary of the southern end of the Grand Canal, it is also one of the major port cities of the ancient Maritime Silk Road. More than 7,000 years ago, Hemudu Culture was born here. In the Spring and Autumn period, Goujian, reigning the State of Yue, built Gouzhang City in present-day Ningbo to demonstrate his hegemony. Later, State of Qin unified the other six states and set up four counties in Eastern Zhejiang, namely Yin County, Mao County, Gouzhang and Yuyao, under the jurisdiction of Kuaiji Prefecture. In the Tang dynasty, Mingzhou was established with its capital seated at the Three-river Junction (the junction of Yongjiang River, Yaojiang River, and Fenghua River), which emerged as one of the four major ports and became the basis of Ningbo's development. In the Northern and the Southern Song dynasties, Mingzhou became one of the three major ports in foreign trade along with Guangzhou and Quanzhou, and achieved unprecedented economic and cultural development. At the beginning of the Ming dynasty, the name "Mingzhou" was changed to "Ningbo" (*lit.* "peaceful waves") with the implication that "The safe sea means peaceful waves". Since then, the name of "Ningbo" has continued to be used to this day. Despite the turmoil in the sea frontiers and the social unrest in the Ming and Qing dynasties, Ningbo was still steadily moving forward in development. With the rise of the Ningbo Commercial Group and the development of the Eastern Zhejiang School, Ningbo's economy and culture were in prosperity again. Later, after First Opium War, Ningbo was forced to open to the world as one of the "five treaty ports". The North Riverbank[①] was designated as a commercial port area and a residence area for foreigners. At the beginning of the 20th century, the North

---

① The North Riverbank, i.e., the north bank of the Three-river Junction, is located in Jiangbei District. It is also the area where the Old Bund is situated. The three rivers are Yongjiang River, Yaojiang River and Fenghua River.

Riverbank gradually became a place where foreigners from all over the world resided, forming the distinctive Old Bund.

In this historical panorama, Ningbo architecture, which reflects the "multifaceted living conditions and political and economic systems of the time and the place" (Liang Sicheng[1]), has also made brilliant achievements. Among them are the stilt architecture at the Hemudu Site of over 7,000 years ago, Pagoda of Tianning Temple and Tuoshan Weir of the Tang dynasty, the Great Hall of Baoguo Temple of the Song dynasty, the Yongfeng Warehouse of the Yuan dynasty, the Tianyi Pavilion of the Ming dynasty, Lin's Residence and Qing'an Guild Hall (or Tianhou Palace) of the Qing dynasty, etc. Looking back on history, there are excellent architectural relics in each period, so that we can touch the history through a stone, a tile, a column or a window. The remains of historical buildings in Ningbo are not isolated examples, but vivid historical stories and a complete picture of interrelated parts. The primitive Hemudu, the ancient Mingzhou, and the modern North Riverbank, all have created a part of the colorful history with Ningbo's regional characteristics through the continuous inheritance and development of Ningbo's architectural culture.

# I. History of Ningbo Architecture

## 1. Ancient Architecture in Ningbo

Ancient Ningbo boasts a wide range of architecture types including primitive architectural sites, urban public buildings, water conservancy facilities, bridges, religious buildings, library buildings, celebrities' former residences, and numerous folk houses.

---

[1] Liang Sicheng (1901–1972), Chinese architect and architectural historian, is also known as the father of modern Chinese architecture.

Archaeological evidence shows that the development of ancient architecture in Ningbo started around six thousand or seven thousand years ago. About that time, Ningbo ancestors started to use the tenon and mortise joinery to build wooden framed houses (such as the stilt architecture of the Hemudu Site). In the primitive settlements, residential areas, burial areas, pottery-making sites, etc. were clearly divided and well arranged. The wooden frame had already taken form, and the plans of the houses were round, square, or shaped like the Chinese character " 吕 " according to different materials and functions. Since the late period of Hemudu Culture, with the development of social productivity, the stilt houses in Ningbo gradually developed into another kind of ground-level building, which became prevalent throughout the feudal era. All this shows that the wooden construction technology in the Yangtze River Basin had reached a fairly high level, obviously higher than that in the Yellow River Basin at that time. The deep historical roots of wooden structures such as Hemudu Culture promoted the emergence of the column-and-tie structure in the later period and inspired the invention and formation of two-story or even multi-story structures, which led to the brilliant achievements of Chinese classical wood structure technology.

During the Qin and Han dynasties, owing to the national unification and prosperity, Chinese ancient architecture reached the first prime in its development. Architectural types such as residences, gardens, villas and city walls developed rapidly. Temples also emerged. The wooden frame of its main structure has almost reached maturity, and bracket sets were widely used in important buildings. The roof began to have diversified forms: The hip roof, the gable-and-hip roof, the overhanging gable roof, and the pyramidal roof all appeared, with some being widely used. Brick-making technology, masonry structure, and arch structure saw new development. There are also some ancient tombs of the Han dynasty in Ningbo, from which many funerary objects unearthed are models of the buildings at that time. Unfortunately, there are no remains of those buildings above ground.

The Western and Eastern Jin dynasties and the Northern and Southern dynasties were a period of great ethnic integration in Chinese history. During this period, traditional architecture continued to develop and Buddhist architecture was introduced. Buddhism, which was introduced into China during the Eastern Han dynasty, began to develop during this period, leading to the extensive building of Buddhist temples and pagodas. According to *The General Annals of Yin County*, in the second year of the Chiwu reign of the State of Wu during the Three Kingdoms period (239), Kan Ze from Gouzhang (present-day Ningbo), Grand Mentor of the crown prince of the Wu, donated his residence to the service of Buddha and had it consecrated as a temple, building the first temple in the history of Ningbo—Puji Temple, with its site located at present-day Cihu Middle School. The "Inscriptions of Rebuilding Puji Temple" of the Ming dynasty mentioned a *futu* (namely stupa) standing in the center of Puji Temple, which is the first Buddhist pagoda in the recorded history of Ningbo. In the third year of the Taikang reign of the Western Jin dynasty (282), the Ashoka Temple in Ningbo was first established, where holy relics of Shakyamuni Buddha, which is extremely precious, are well preserved in a stupa, the smallest ancient pagoda in China. Moreover, Tiantong Temple in Ningbo was built in the first year of the Yongkang reign of the Western Jin dynasty (300), where most of the existing buildings were rebuilt in the Chongzhen reign of the Ming dynasty.

There are no extant examples of ground buildings from the Eastern Han dynasty to the Six Dynasties period. However, the sporadic literature records and archaeological evidence from ancient tombs in Ningbo, which includes the pottery funerary objects of house models and jars with watchtowers and figures, vividly reproduce the architectural forms during that period, including single buildings and building clusters. At that time, the Ningbo folk houses were square, rectangular or in the shape of Chinese character "日" in plan view. The gate was designed in the middle of one wall of the house, or off to the side. With a few load-bearing walls, a house was mainly constructed with wood-framed structures. With the maturity of column-beam-and-strut frameworks and column-and-tie-beam frameworks, the technology of wood-framed structures increasingly improved, leading to a considerable growth of multi-story buildings in number. The walls that used to be built with rammed earth were made with stacked bricks in the later stage. Barrel tiles and eaves-end tiles were used on the roofs, most of which were overhanging gable roofs, hip roofs, pyramidal roofs and gable-and-hip roofs.

To sum up, the architectural style of this period can be described as robust, rough and slightly immature at first, but in the later period, it has shown a trend of being vigorous, beautiful, sturdy and slightly soft. This is a thriving stage of development in the history of architectural style in Ningbo, showing the cultural characteristics and social progress of the region and the period.

The architecture of the Sui and Tang dynasties not only inherited the achievements of the previous dynasties, but was also mixed with the foreign styles, forming an independent and complete architectural system, leading the ancient Chinese architecture to a mature stage, and exerting an influence over the Korean Peninsula and Japan. Architecture of the Tang dynasty kept in Ningbo include Tianfeng Pagoda, Pagoda of Tianning Temple, Tuoshan Weir, and two Dhanari columns moved to and preserved in Baoguo Temple in 1984. Among them, Pagoda of Tianning Temple is the only existing multi-eaved brick pagoda in the Tang dynasty that has been completely preserved in Jiangnan (lower reaches of the Yangtze River) of China, and it is also one of China's early examples of the form of twin pagodas in front of the temple.

Since the Tang dynasty, with the development of the port city of Mingzhou, many water conservancy projects have been built in Ningbo. Among them, Tuoshan Weir, known as one of the four major water conservancy projects in ancient China, is the most famous. Tuoshan Weir construction is an extremely important guarantee for the water supply and overall development of Mingzhou port-city, and still plays an important role today.

The Dhanari columns of Puji Temple in the Tang dynasty are the biggest and oldest ones in Zhejiang Province. This structure illustrates the architectural form and specifications in the fourth year of the Kaicheng reign of the Tang dynasty (839).

In the second year of the Chongning reign of the Northern Song dynasty (1103), the imperial court issued and published *Building Standards* (*Yingzao Fashi*). This is a standard manual on architectural design and construction, and a perfect technical treatise on architecture. It was published for better management of the construction of official-style architecture such as palaces, temples, government offices, and official residences. This book reflects that in the Song dynasty, the ancient Chinese architecture reached a new historical level in engineering technology and construction management, and enjoyed a unique position in the history of world architecture. The Great Hall of Baoguo Temple in Ningbo, built in the sixth year of the Dazhong Xiangfu reign of the Northern Song dynasty (1013), is the earliest completely preserved wooden structure in Jiangnan. After various measurements and long-term studies by experts, we now know that its structure is consistent with the architectural theories and standards described in *Building Standards*. It shows that practice tends to precede theory. Baoguo Temple is one of the first batch of the Major Historical and Cultural Sites Protected at the National Level, which includes only three in Zhejiang Province and more than 80 across the country. We can see its important and unique position in the history of Chinese architecture.

The memorial arches in China were all made in wood before the Southern Song dynasty. The stone arch built in the second year of the Shaoxing reign of the Southern Song dynasty (1132) in Hengsheng Village, Yinzhou District, Ningbo, another of the Major Historical and Cultural Sites Protected at the National Level, is an imitation of the wooden frame typical in the Southern Song dynasty. It not only displays the characteristics of the wooden arch, but also possesses obvious features of the transition from wood to stone arches, providing us with first-hand materials to study and authenticate the stone arches in the Southern Song dynasty.

In Ningbo area, the typical bridges in the Song dynasty are Xi'ao Stone Arch Bridges in Ninghai County. Their advanced engineering serves as a valuable example for us to study the bridge construction technology in Eastern Zhejiang.

Xi'ao Stone Arch Bridges in Ninghai are composed of Huide Bridge, Citang Bridge and Siqian Bridge. The three bridges, all single-arch stone arch bridges, run from east to west across the same stream. They are the only ancient stone arch bridges in such form and with such arch characteristics in Zhejiang Province, and therefore are very rare.

The Yuan, Ming and Qing dynasties ruled China for over 600 years. There are very few architectural remains from the Yuan dynasty because it lasted only about a hundred years. In 2002, a large-scale architectural site of the Yuan dynasty was discovered during the archaeological excavation of the Zicheng City (inner city) Site near the Drum Tower. After expert evaluation, it was determined that this architectural site is the remains of Yongfeng Warehouse of the Yuan dynasty, which is the largest single architectural site of the Yuan dynasty found in China so far and is unique across the country. This discovery of the Yongfeng Warehouse Site of the Yuan dynasty filled a gap in the archaeology of the Yuan dynasty in China and was rated as one of the top ten national new archaeological discoveries for 2002. The other remaining architectural structures of the Yuan dynasty are the West Pagoda of Ashoka Temple and Guangji Bridge in Fenghua.

There are a number of buildings of the Ming dynasty kept in Ningbo, among which the most famous is Tianyi Pavilion (*lit.* heaven one pavilion). Its name originated from the saying "Heaven one and Earth six"[1]. As a unique building in Ningbo, it is the oldest existing private library in China, with a history of more than 400 years. Among other buildings, the most typical is the architectural complex of the Ming dynasty in Cicheng, Ningbo, with Jiadi Shijia Residence (*lit.* residence of a family with several *jinshi*, scholars who passed the highest imperial examination) and Danai Hall (*lit.* Hall of Great Tolerance) as the representatives. The hall of Dafang Yuedi Residence at the Moon Lake is so magnificently grand that it is very rare in Ningbo. Fan's Residence is the largest and best-preserved residential building of the Ming

---

[1] According to the River Diagram in Taoism, Heaven one generates water; Earth six forms water.

dynasty in Ningbo, characterized by a strict architectural structure, stout materials and a simple design.

The Qing dynasty saw even more buildings. They generally used smaller materials but paid great attention to decoration. For example, Lin's Residence is three bays deep, shows a symmetrical layout with several axes, and possesses multiple spare lanes; its size is grand and its overall layout follows no rigid standards. The buildings have the following characteristics: Firstly, a frequent use of *goulian daqian* structure (connected beam frame structure, i.e. the beams are connected with each other, so that the roofs are connected too), tall and thin king posts, with the lower joint resembling an eagle's beak or a lotus-leaf stalk; secondly, the *yueliang* beams (*lit.* crescent-shaped beams) on windowed verandas are generally carved with various patterns; thirdly, the columns are cylindrical in shape, some covered with wood planks; fourthly, the column bases are either drum-shaped or in the shape of a tall felt cap. In the late Qing dynasty, the structure became much more complicated, which involves exquisite wood carving, stone carving and brick carving. The windowed verandas are narrow in span, but luxurious, ornate and complicated with *hualan* beams (cross-shaped beams) and *queti* (beak-shaped braces) full of carvings. The bracket set is in relatively small proportion to the building, and the intercolumnar bracket set is generally placed between intermediate purlins and intermediate tie-beams, for decoration purposes only. In some buildings, the bracket sets are so exquisitely carved that they provide a sense of intricacy, complexity and luxury.

Some other famous ancient buildings include Tiantong Temple, Ashoka Temple, Ninghai Ancient Stage, and Qing'an Guild Hall. They are also listed as the Major Historical and Cultural Sites Protected at the National Level which represent the ancient architecture in Ningbo.

Ashoka Temple, named after Ashoka King of India, is one of the most famous temples in the country with its spectacular layout and construction. It covers an area of 124,000 square meters with a building area of 23,400 square meters. It has been a famous ancient temple in the East Asian cultural circle since the Song dynasty, possessing a close relationship especially with the Japanese Buddhist culture.

Tiantong Temple, a famous temple in Jiangnan, is also one of the famous temples across the country. Covering an area of 76,400 square meters, the building complex is not only large in scale, but also exquisite in design. Since the Song dynasty, Tiantong Temple has not only been the source of the transmission of Buddhism in East Asia, but also an important source of the export of architectural culture. Rinzai Zen and Soto Zen of Japanese Buddhism both regarded Tiantong Temple as their ancestral temple, and have built a large number of Zen temples in imitation of the architectural pattern of Tiantong Temple. For example, the Daihongzan Eiheiji Temple of Soto Zen in Japan followed the pattern of Tiantong Temple in the Song dynasty, so it is also known as "Little Tiantong Temple".

The self-contained Chinese wooden frame architecture has had a profound impact on the architectural culture of surrounding countries. During the Tang and Song dynasties, Japan, Koryŏ and other countries sent craftsmen to China to learn the construction technology of wooden frames, and many skilled craftsmen in China were also invited to sail across the sea to those countries to build temples and gardens for them. As an important hub for foreign trade and cultural exchange, Ningbo has played an important role in the dissemination of architectural culture. For example, the eminent Japanese monk Chogen came to China three times since the fourth year of the reign of Qiandao of the Southern Song dynasty (1168) to study Buddhism, Chinese calligraphy and other traditional culture. During his stay in Ashoka Temple, he used a large amount of wood transported from Japan to help build the Sarira Hall there, and learned the construction skills from it. When he returned home, he invited a large number of carpenters from Mingzhou and other places to Japan to rebuild the famous Todaiji Temple. Eisai, another Japanese learned monk, helped build the Qianfo Pavilion (thousand-Buddha pavilion) in Tiantong Temple when he studied Zen there. He learned a lot of architectural construction techniques through studying and practicing. Later, he built a number of famous temples with Chinese traditional style in Kyoto, Kamakura and other places in Japan, which further spread traditional Chinese architectural culture in Japan. At the beginning of the Qing dynasty, Zhu Shunshui, a well-known scholar, went into exile in Japan after the failure of the anti-Qing movement. Having brought to Japan books and materials on the construction techniques and styles of Jiangnan gardens, he designed and built Korakuen Garden in Tokyo.

While China's traditional architectural culture brought about a profound influence on the architectural styles and urban views of surrounding countries, it also absorbed and integrated the cultural nutrition from other countries into its own traditions. For example, the introduction of Buddhism and the construction of pagodas, temples and other Buddhist buildings have enriched the forms of ancient buildings in China. As Ningbo has been one of the major port cities of the ancient Maritime Silk Road and an important port-city for foreign trade on the southeast coast, the influence of foreign culture is clearly presented in its local architectural style, as evidenced by such ancient buildings as Tianhou Palace and the Moon Lake Mosque. The integration of foreign architectural styles produced an unprecedented power in modern times, bringing a new look to Ningbo's architectural culture.

## 2. The Modern Architecture in Ningbo

After the First Opium War, the Qing government succumbed to the force of the British invaders and signed the humiliating Treaty of Nanking. In the following year, it signed the General Regulations of Trade Between Britain and China. Since then, as one of the first treaty ports, Ningbo port was forced to open to Western powers. According to the provisions of the relevant treaties, "On the twenty-eighth day of the tenth lunar month, the twenty-third year of the Daoguang reign (1843), foreign ships, one big, one small, and one warship, led by the chief Robert Thom, sailed to Ningbo port... On the twelfth day of the eleventh month of the same year (the New Year's Day of 1844), all the civil officials and military officers were invited to watch the opening of the Ningbo port..."[1] Thus, Ningbo port officially opened. Subsequently, 12 countries including Britain, France and the United States set up their own consulates along the Old Bund on the north bank of the Three-river Junction.

At the beginning of the port opening, the countries that came to Ningbo for trade included Britain, France, the United States, Germany, Russia, Spain, Portugal, Sweden, Norway, and the Netherlands. According to the rules at that time, "After the opening of the five ports, the British businessmen (and other foreign businessmen) are only allowed to trade in the five ports, but not allowed to go to other ports, nor are Chinese people allowed to collude with any foreign businessmen in private trade in other ports."[2] Therefore, Britain and other Western countries established a stronghold in Ningbo to control the foreign trade and economic lifeline of Ningbo port. In 1850, they forcibly designated a large area along the north bank of the Three-river Junction as "the foreign residential area and trading port area", which gradually became a bridgehead for reinforcing the Western control over Ningbo port. According to the archives of the British Consulate, the invaders intended to include Zhenhai and the southeast townships of Yin County into their settlements, which covered "the land along Yongjiang River to the east, and the area outside the city gate extending to Dongqian Lake to the south, Banpu and Liang Shanbo Temple to the west, and Zhenhai Estuary to the north"[3]. Due to the resolute opposition of the people of Ningbo, this plot failed.

What the Western ships brought was the embryo of Ningbo's modern city. After the opening of the port, foreign businessmen set up firms in Ningbo. In foreigners' residences at the North Riverbank, Western buildings of consulates, banks, churches, customhouses, and police stations were built one after another, and the place gradually developed into *yangchang* (a big

---

①   Cited from *Foreign Affairs in Their Entirety: The Daoguang Reign* (Vol. 69).
②   Cited from *A Collection of Old Treaties Between China and Foreign Countries* (Vol. 1), Wang Tieya.
③   Cited from "The origin of Ningbo port and the development of modern trading port zone", in the *Bulletin of the Institute of Modern History, Academia Sinica* (Issue 2), Wang Ermin.

city infested with foreigners), inhabited by people from different countries. For example, the first cement road in Ningbo ran along the North Riverbank, and the earliest street lights were set up there. The tallest building then in the city, the business building of the Ningbo branch of the Imperial Bank of China, was erected there. Ningbo's first telephone company was established there. As the first railway station in Ningbo was also built there and open to traffic in 1914, the state-owned railway station financed by private capital is the earliest railway station for both passengers and freight transportation between Xiaoshan and Ningbo, whose Ningbo section connected Mazhu Station in the west and the old Ningbo Station in the east. The present-day Daqing Road was originally on the railway line. After the Chinese People's Nationwide War of Resistance Against Japanese Aggression began, the railway was ordered to be demolished in order to thwart the attack of Japanese army. Still, owing to this railway, Ningbo became a transportation hub in Eastern Zhejiang in the period of modern history (1840–1949) and witnessed its gradual change from a feudal city to a modern city and later to one of the main commercial centers in Eastern Zhejiang. As a result, Ningbo formed a layout of the ancient city in the south and the commercial port area in the north. The architectural style along the Old Bund is characterized by a combination of Chinese and Western styles. As a center for a new lifestyle, a mixture of the elegant East and the romantic West, the Old Bund presented a new cultural look different from the traditional Chinese society and became an epitome of the modern Ningbo port-city.

The development of Ningbo City and its architecture in modern times is different from that of the major modern cities in China, such as Shanghai, Tianjin and Wuhan, which had relatively complete foreign settlements and careful urban planning. It is also unlike that of modern leased land and railway-affiliated cities in China, such as Harbin, Dalian and Qingdao, as Ningbo's urban construction was based on the urban planning model prevailing in Western countries at that time and was basically affected by the surrounding cities (mainly Shanghai). However, Ningbo's modern architecture did not blindly imitate Western architecture. In the sense of integration of Chinese and

Western architectural cultures, it displays more flexibility, more freedom and more folk creativity, thus showing its unique architectural style.

The development of modern architecture in Ningbo can be roughly divided into four stages from emergence to prosperity and to decline.

### The Period of Emergence (before 1842)

According to *The General Annals of Yin County*, the following were the earliest modern buildings in Ningbo: A Catholic church built by the Italian missionary Martino Martini in the fifth year of the Shunzhi reign of the Qing dynasty (1648) (the address unknown), a small church and some residences on the Yaohang Street built by the French Protestant missionary Jean-Alexis de Gollet in the fortieth year of the Kangxi reign of the Qing dynasty (1701), all of which were later destroyed. This is the beginning of Ningbo's modern architecture.

### The Period of Preliminary Development (1842–1895)

This period is the early stage of the development of modern architecture in Ningbo. China's transformation from a traditional agricultural society to a modern industrial society brought about the emergence of modern architectural types, technologies and forms. However, this transformation is not the result of China's natural social development, but the passive sudden transformation caused by the invasion of imperialism. As one of the earliest treaty ports, Ningbo gradually formed a new urban area that was independent of the old urban area and centered around foreigners' settlements, namely the Old Bund at the Three-river Junction. After Ningbo opened its port on January 1, 1844, modern architecture with foreign characteristics began to flourish. Britain, France, and the United States built consulates in Ningbo, and Catholics built churches and church schools; various municipal administrative bodies were also set up, such as Zhejiang Customs and Jiangbei Police Station. These buildings were mainly designed and built by foreigners. From the exterior appearance and floor plan to the internal structure and decoration, they are basically of Western style or Western colonial style. Most of these early modern buildings were built along the west bank of the Yongjiang River (i.e., the area along present-day Xinmalu Road). They all faced east onto the Yongjiang River. A space of more than 30 meters in width was left between the buildings and the river bank for the purpose of loading and unloading the goods on the ship and allowing the passage of civilian boat trackers.

These buildings gradually altered the view of the Old Bund and turned Ningbo into a Chinese city with a touch of Western style. However, it didn't significantly change Ningbo's traditional architectural system. The early modern buildings that have survived to date mainly include Zhejiang Customs built in 1861, Jiangbei Police Station built in 1864, Jiangbei Catholic Church built in 1872 and the additional bell tower built in 1899, Swire Group built in 1879, and the British Consulate built in 1880.

### The Period of Full Development (1895–1937)

In the process of cultural exchange between China and foreign countries, the development of modern architecture in Ningbo entered its third stage, which is mainly characterized by the contact, coexistence and mutual influence between its tradition and the foreign architectural art. Due to the different attitudes of people from different social strata towards traditional and the Western culture, different fusion techniques were used in the architectural design, which enriched Ningbo's modern architecture. At the same time, the new building materials, structures, construction technologies and equipment were also gradually introduced from abroad. After the 1911 Revolution, many Chinese architects, engineers and construction businessmen studying in Western countries returned to China to set up architectural design and construction factories. By combining the advantages of Western architecture with the local traditional architectural style, they built a number of modern public buildings, Western-style houses and residential houses, whose styles include Western classicism, eclecticism, a combination of Chinese and Western styles, and traditional wooden frame with Western decorations.

After 1900, due to the imperialist occupation of the area along Waima Road the in the North Riverbank, except for the buildings owned by a handful of wealthy capitalists, most buildings were built in the hinterland of the North Riverbank, including the area to the west of present-day Xinma Road, Zhongma Road and People's Road.

The main buildings in this stage include the former sites of the Water Headquarters of the Japanese invaders, the Ningbo branch of the Bank of China, Ningbo Post Office, the Ningbo branch of the Imperial Bank of China, among other public buildings, and a number of *shikumen* building complexes as well.

The former site of the Water Headquarters of the Japanese invaders was originally Xie Hengchuang's private residence built in 1905. It is a three-story building made of cement bricks, which adopts Western architectural style and is decorated with Chinese traditional auspicious patterns, such as "Five Bats Holding a Longevity Peach" (symbolizing happiness and longevity) and "Three Promotions in a Row". It's quite impressive that each room is equipped with a Western-style fireplace and a chandelier. After the fall of Ningbo, the house was occupied by the Japanese invaders and soon became the "Water Headquarters" with checkpoints set up there.

The Bank of China also had a branch in Ningbo. According to a survey report in 1943 by Zhejiang Industrial Bank, Shanghai ranked first in national commercial capital, bank capital ranked first in Shanghai commercial capital, and Ningbo ranked first in bank capital ownership. The Bank of China was established in 1912. It took over the bank business of the Qing dynasty and became the national bank. Its head office was located in Shanghai with a capital of 60 million yuan.

The former site of Ningbo Post Office, built in 1927, is located at No. 172, Zhongma Road. The five-bay, two-story building is built with a combination of black bricks and red bricks. Its veranda is of multi-arch structure of Western style. The post service in Ningbo began in the fourth year of the Guangxu reign of the Qing dynasty (1878). In the twenty-third year of the Guangxu reign of the Qing dynasty (1897), with the establishment of the Great Qing Post, the Post Office of Ningbo Postal District was set up. In the second year of the Xuantong reign of the Qing dynasty (1910), the General Post Office of Zhejiang Postal District was established and Ningbo became a secondary postal district, so the name was changed into the Secondary General Post Office. After the 1911 Revolution, the Great Qing Post was renamed Chunghwa Post. In January 1914, the Ningbo branch of the Chunghwa Post Office was changed into Yinxian First Class Post Office. In 1927, it was changed into Ningbo First Class Post Office. In 1931, it was renamed Yinxian Post Office. In 1947, it was moved to the new office building at No. 122, Chezhan Road.

The former site of Ningbo branch of the Imperial Bank of China, which was originally built in 1930, was located at No. 29, Waima Road, and was divided into east and west parts. The main building is a five-story building. In the west part are three-story single-eave buildings with corridors in the middle. The front gate is decorated with black marble in the shape of a pagoda. The second gate uses white marble for decoration. In the interior, the ceiling and surrounding walls are decorated with plaster, which is characteristic of foreign architecture. In the twenty-third year of the reign of Guangxu of the Qing dynasty (1897), Ningbo people who resided in Shanghai initiated the establishment of the first bank in China's history, the Imperial Bank of China, and set up a branch in Ningbo in 1921.

**The Period of Decline (1937–1949)**

In 1937, the Chinese People's Nationwide War of Resistance Against Japanese Aggression broke out, which caused the sudden suspension of the booming modern Chinese architecture. Since then, the modern Chinese architecture entered its withering period, with many cities and buildings destroyed by war, and the overall construction activity brought to a halt. After the victory of the Chinese People's War of Resistance Against Japanese Aggression in 1945, the Civil War throughout China broke out, and the suspension of construction activities continued until 1949. Despite a small amount of construction in Ningbo during this period, it was nothing compared to the damage of the war, and few buildings of that period have survived till today.

The surviving outstanding modern buildings in Ningbo are the physical witness of the historical development of modern Ningbo. They emerged from the invasion of imperialist powers, but the buildings themselves embody the wisdom of the people. With their exquisite artistry reflecting the integration of Chinese and Western cultures, they are undoubtedly a valuable cultural heritage of the city.

# II. The Cultural Significance of Ningbo Architecture

Architecture is an important representative of a local culture of an era. Culture affects the style of architecture. From architecture, we can see the culture at a specific time and place. Ningbo is located in the east of Ning-Shao Plain in Eastern Zhejiang, close to the East China Sea, facing the Zhoushan Islands across the sea in the east, bordering Hangzhou Bay in the north, and connecting Taizhou in the south. There are two main mountains in Ningbo, Siming Mountain and Tiantai Mountain, and three rivers, Yongjiang River, Yaojiang River and Fenghua River. The unique natural environment has formed the unique culture with the regional characteristics of Ningbo. The traditional culture of Ningbo is characterized by the combination of scholar culture and merchant culture. The scholar families of Ningbo passed down their agriculture and scholarship as the family tradition, and wealthy merchants established themselves through setting up businesses. This is the cultural basis of Ningbo architecture. To be specific, the cultural connotation of Ningbo architecture lies in the following aspects.

## 1. A Clan Living Together and Paying Great Care to *Fengshui*

The traditional village in Eastern Zhejiang is a typical place where a clan lived together. The concept of respecting and obeying the rites and rituals is directly reflected in the residential houses. One of the most important characteristics of patriarchal society is the ancestral hall system. Ningbo ancestral halls generally adopt a strict symmetrical layout, and the courtyard space is composed of several sections of buildings (one section means a courtyard with one entrance gate), generally including the gate, *yimen gate* (*lit.* gate of rites), the main hall, and the back bedroom in succession. The back bedroom is the place where the memorial tablets of ancestors are placed and the portraits of ancestors are hung, which is exactly what Wan Sida, a Confucian scholar from Ningbo in the early Qing dynasty, meant by "people of the same clan are led to offer sacrifice for the sake of worshiping their ancestors and uniting the surviving families" in Chapter "Patriarchal Law" in *The Inquiry into the Rituals.* Ningbo has a good preservation of the ancient ancestral halls along with a complete set of related ethical systems, such as Qin's Ancestral Hall in Haishu District, Xie's Ancestral Hall of the First Ancestor in Simen, Yuyao (a county of Ningbo), and Sunjiajing Ancestral Hall in Cixi (a county of Ningbo). Hanling Village, a famous historical and cultural village at the municipal level, has clans of several different surnames living together. The major surnames with more than 100 households can have their own ancestral hall built or *tangqian* (*lit.* in front of the main hall) set up. Other surnames with more than 10 households can have their own *tangqian* established. The other non-settled surnames can also have their *tangqian* built after joining the Juxinghui Association. Today, in Hanling Village are kept Jin's Ancestral Hall, Sun's Siben Hall (*lit.* to keep in mind the roots of family), Zheng's Chongde Hall (*lit.* to advocate morality), Kong's Tangqian, Ling's Tangqian, Zhou's Tangqian and so on, which provide the evidence of the clan culture of living together.

Among the many beliefs that have played a role in the development of Ningbo's architectural culture, the belief in "harmony between humanity and nature" in the *fengshui* (*lit.* wind and water, or geomancy) system is the most fundamental one, which emphasizes the inseparable relationship between architecture and a harmonious coexistence between humanity and nature. The arrangements of traditional folk houses in Ningbo pay attention to the relationship between different priorities, levels and sequences and conform to *fengshui*, and therefore almost all of them display the "harmony between humanity and nature" and reflect the house owner's social status, economic power, personal interests, family structure, customs, religious beliefs, mode of production and lifestyle.

Great importance was attached to the *fengshui* concept of "harmony between humanity and nature" in the sitting of many ancient villages in Ningbo. It is believed that *fengshui* is related to the rise and fall of villages and clans. "The *fengshui* master, or the soothsayer who divines the house, finds out the favorable and unfavorable signs of the land. The good signs can bring peaceful spirits and prosperous future for their future generations, just as a plant that takes its deep root is likely to flourish." This is basically consistent with the idea of *fengshui*, which emphasizes the interaction between humans and the environment, pursues harmony with heaven, earth and all beings in nature, and aims at winning good fortune and avoiding evilness. Folk houses are the most important component of a village. Their orientation, form, layout and inter-relation are affected by the concept of *fengshui*. In Ningbo folk houses, *fengshui* is mainly reflected in meeting, avoiding, amulets and charms in architectural designs. The vast majority of traditional houses are designed in a regular and square courtyard layout, which agrees with the *fengshui* concept that "the house should be laid out in an orderly fashion, without any sharp angle, obliqueness or skewness". The door has special spiritual significance in *fengshui*. Generally, the door is arranged facing east, as it is believed that the "purple air", an auspicious omen, comes from the east.

The buildings in Ningbo basically face south or east. From the cultural perspective, the orientation of folk houses is related to the idea of "ruling the world by facing the bright side". Chapter "Explaining the Hexagrams" in *The*

*Book of Changes* says that "Facing the south, the sage listens and governs the world by facing the bright side". Confucius also mentioned in Chapter "Yongye" in *The Analects*: "Ran Yong can be given a position facing south." "Ruling the world by facing the bright side" is actually "ruling the world by facing the sun". This is the unique culture of "facing the south" in ancient China. Since most of the time the sunlight comes from the south, and people's life and production are based on obtaining sufficient sunlight, people naturally tend to face south for lighting. Over time, people's *fengshui* concept of "living facing the south" came into being.

### 2. The Courtyard Space Designed Uniquely According to the Local Conditions

It is relatively easy to build a house, but quite challenging to build a community with superior environmental facilities. It requires long-term investment and construction, often demanding the continuous efforts of several generations. Instead of following the stereotyped ways in their construction, people in Ningbo handle the overall layout and various complex issues with flexible methods according to different situations, showing strong adaptability.

If the foundation of the chosen site is square or rectangular, it is easy to arrange the layout of the building, but due to various reasons, the foundation is often irregular-shaped in actual construction, which requires careful planning according to the local conditions in the actual construction. The Yanshou Hall in the west part of Moon Lake is a case in point. Its foundation has a very irregular shape, but its proper design and arrangement make it exquisite. The residential area in Huizheng Lane has a trapezoid-shaped foundation and some twists and turns, but the relation between the buildings is decently and impressively dealt with. Hu's Residence in Lianqiao Street has a narrow and long foundation, but with careful design, the houses are delicately connected and amazingly organized. Along the Old Bund in Jiangbei District, many modern buildings facing the street were designed to fit in with the winding road and therefore achieve harmony of the whole street.

Building materials were also chosen based on the local resources. For example, Xujiashan Village in Ninghai took advantage of the abundance of local rocks to build walls with it and is therefore called "Rock Village".

The courtyard space of buildings in Ningbo is quite distinctive, with the surrounding houses connected around a courtyard, or *tianjing* courtyard (*lit.* sky well). It has the following characteristics.

First, the courtyard has a *tianjing*-style layout, which is closely associated with the geographical environment and climate of Jiangnan. Jiangnan is densely populated, with many hills and little arable land, which makes it necessary to save the land as much as possible when building houses. Therefore, houses were built on three or four sides of a courtyard. At the same time, due to the hot and humid summer and the cold and bleak winter in Jiangnan, people built two-story buildings surrounding a tall and narrow *tianjing* on three or four sides. It is meant for better air movement and the effect of more warmth in winter and more coolness in summer.

Second, *tianjing* is also designed to let out foul air from the house, improve indoor lighting, and collect rainwater, which will be drained through the gutter. The rainwater runs over the roofs before being collected and then let out on the ground, implying that the rainwater, a symbol of fortune in Chinese culture, flows from four sides into the courtyard and that good water does not flow out.

Third, high walls and narrow alleys were designed. In order to prevent the spread of fire between the courtyards, the gable walls have a ladder-shaped contour and are built higher than the roof, so they are named fire-resistance gable walls. Such *tianjing*-style courtyards are closely connected to each other by alleys, which also serve as the paths connecting the alleys, streets, and main roads in the village. All these alleys make a transportation network resembling a fishbone. In order to save the land, streets and alleys are usually narrow, with the space of the alleys defined with gable walls. The high walls and narrow alleys have become the typical form of Ningbo folk houses.

### 3. Unique Craftsmanship in Rich and Unique Architectural Decoration Art

The decorative art of folk houses in Ningbo is ingenious and distinctive, and displays a high artistic level, whether they are "three carvings" (wood carving, brick carving and stone carving), or mural paintings and color paintings. The high-quality brick carvings are mostly seen on gatehouses. The stone carvings are mostly used for stone windows, stone drums, *baogushi* (drum-shaped gate bearing stone pillar), balustrade capitals, balustrade panels, *shique* (stone tower erected in front of palaces, temples and tombs to commemorate the owner's honors and achievements), *queti*, and so on. The wood carvings are mostly used for bracket sets, sealing boards, balustrade panels, wooden windows, wooden gates, wooden doors, and column tops. Among the decorative art in Ningbo, wood carving is the most prominent.

Wood carvings are widely used in traditional residential buildings in Ningbo. In the large wooden structures ranging from beam frames, purlins, bracket sets, *tuofeng* (a camel's hump-shaped brace) and some small wood components including gates, doors, windows, balustrades, *niutui* (a wooden component like corbel, which sticks out from the top of column to support the horizontal structural elements above it) and *queti,* you can see exquisite, gorgeous and three-dimensional carvings, complex but well designed. The carving patterns feature flowers and other plants, birds, animals, human figures, landscapes and geometric patterns. Among the wood carvings, the most noteworthy one is the red-lacquered wood carving covered with gold foil, which was listed in the first batch of national intangible cultural heritage items in early 2006. It has a history of more than 1,000 years, originating from the art of carved painting and gold foil decal in the Han dynasty, a decorative architectural wood carving with both color paint and gold foil.

The brick and stone carvings of Ningbo folk houses are delicate, vivid and of high artistic value. Those on the gatehouse are mostly based on historical stories, such as "Eight Immortals Crossing the Sea", "Jiang Taigong Fishing", and "Jiangxianghe" (conflict resolution between General Lian Po and Prime

Minister Lin Xiangru). Displaying skilled carving crafts, Lin's Residence at the Moon Lake is the representative of brick carving in the residential buildings of the late Qing dynasty in Eastern Zhejiang. The carving patterns include "Goddesses Present Flowers", "Luan and Phoenixes Sing Together" (a symbol of happy marriage), "Literati Gathering", "Magpies on the Tree" (meaning happy faces), "Harmonious Family", "Respecting the Elderly and Loving the Young", "Living Together Till Old and Grey", "Gold and Jade Fill the Hall", "Rising in the Official Rank", "Heavenly Officials Sending Blesses", "Magu the Longevity Goddess Extending Birthday Wishes", among others. These patterns in the courtyard house not only carry the good wishes of its residents, but also represent their eternal pursuit of a better living environment for future generations.

## 4. Keeping a Low Profile and Focusing on Pragmatism

The large houses preserved in Ningbo are generally encircled by high walls, with the main gate to the side, simple and plain. After Ningbo was opened as a trading port in the late Qing dynasty, many people started doing business and tended to avoid showing off their wealth. As a result, the owners of the houses insisted on the implication of hidden depth in the design of their residences. For example, the gate of the Qing'an Guild Hall in Ningbo is a medium-sized brick gatehouse. It suggests that the owner did not want to show off with it. The "unfathomable" and "hidden" (i.e., hide one's light under a bushel) is also the philosophy of Chinese scholars and bureaucrats. Take Lin's Residence at Zijin Street for example. Its extraordinarily small main gate and the simple carvings in the screen wall facing the gate form a strong contrast with the complicated brick carvings in the screen wall of the inner courtyard. Take another example—the Former Residence of Yang Fang. The decoration of the outside of the *yimen* gate is simple and plain, while that of the inside is complex and intricate. Even the two Chinese characters " 杨坊 " (Yang Fang) on the lintel are carved inside rather than outside, which can be described as another good example of low-profile and reserved design.

In terms of materials used for traditional folk houses in Ningbo, people would mainly choose the wood of the dominant local trees such as fir and pine, and the dominant local rocks such as black rock and Meiyuan rock. The rare wood is relatively scarce in folk houses. Most of the residential buildings are not gorgeously decorated, but rather practical and elegant, without many carvings. Generally, the carvings are mainly designed on *baotou* beams (*lit.* beam holding the head, or the beam connecting an eave column and an interior column), *niutui*, *queti* and column bases. Mostly the indoor beam frames are not carved. Columns and beam frames are usually light and practical, which form a strong contrast with Huizhou architecture.

The color of exterior walls is devoted to a plain and natural aesthetic view, with cold gray as the main tone, and black and white as the basic color. The black bricks, white walls, or plain brick walls, and black tiles form a simple but unified architectural color with the gradation of black, white, and gray. The hollow plain wall, solid wall, and tile wall present the beauty of simplicity and elegance. It is said that this style of color is influenced by Zhu Xi, a neo-Confucian of the Southern Song dynasty, who proposed that "Sages tend to use plain language which contains deep and profound messages".

## 5. The Combination of Chinese and Western Styles, with Newness and Diversity

After the First Opium War, Ningbo became a treaty port, which led to the emergence of the modern Western-style architecture characterized by a unique combination of Chinese and Western elements. This architectural style represents an important type as well as a significant stage in the history of modern Chinese architecture.

Since their emergence, most modern residential buildings in Ningbo have been marked with a coexistence of Chinese and Western styles. Seen from the early *lilong* (alleys) houses, those modern houses still contain the characteristics of traditional Chinese houses, but their overall layout originates from Europe. From *shikumen*, the most distinctive architecture of the early *lilong* houses, we can easily find the coexistence of Chinese and Western cultures. The inner structure of *shikumen* houses is derived from the traditional Chinese residential courtyards, but the overall layout adopts the Western town house style, which is based on the consideration of land use efficiency to meet the social situation at that time; its door frame, black gates and copper door ring are all characteristic of traditional Chinese architecture, while the triangular or arc-shaped lintel decoration showcases the Western style. More importantly, this type of architecture does not belong to any traditional Chinese residential architecture, nor is it an imitation of any kind of Western architecture. They are unique modern folk houses specific to Ningbo with a combination of Chinese and Western architectural characteristics.

Despite the complicated political and economic background against which Ningbo was involuntarily opened, it's a big chance for architects to introduce a variety of architectural designs popular in the West at that time, resulting in a constant change in the architectural style in Ningbo. At the beginning of the opening of Ningbo port, many Western-style buildings were built along the Old Bund. By the end of the 19th century and the wake of the 20th century, with the introduction of new ideas and development of new materials, Ningbo architects had already well integrated the Western residential culture and the local residential concepts into their architectural design. For example, the Catholic Church with Gothic architectural style, the British Consulate and Zhejiang Customs with the style of British colonial-style architecture, the Police Station blending Chinese structure with Roman columns on the facade, the British Business Firm with Western-style Roman columns, the Western-style garden house owned by Yan Zijun, the son of Yan Xinhou, and the Western-style garden house decorated with Corinthian columns owned by a Ningbo businessman Zhou Jinbiao, etc.

The pursuit of new trends is also a prominent social mentality in modern Ningbo with a developed market-oriented economy. The idea that "Newness is beauty" is reflected in all aspects of modern society, including drama, literature, clothing and architecture. On May 25, 1936, Lingqiao Bridge in Ningbo was completed and open to traffic. It is a rainbow-shaped, silver gray steel structure bridge with red-colored bars over Fenghua River. The long-cherished wish of Ningbo people was finally realized. Mountains of people came to take a look at it. Ningbo people's special willingness to embrace new things is well evidenced by their constant pursuit of newness in architecture.

# III. The Cultural Value of Ningbo Architecture

The historical architecture of Ningbo are not only the witness of its history and culture, but also an encyclopedia for us to understand Ningbo's history and culture, which are worth cherishing and preservation. Mr. Liang Sicheng, a famous architectural historian in China, pointed out that the most effective protection of ancient architecture is to inform people of its value. Only when everyone is aware of its value can they consciously protect it. In order to have the historical buildings in Ningbo effectively protected and the historical architectural culture inherited, the first priority is to improve the public's awareness of the cultural value of the historical architecture.

## 1. Historical Value: Preserving Unique Historical Buildings

Ningbo's historical architecture has outstanding achievements and unique styles, and plays a significant role in the history of Chinese architecture and even in the history of world architecture. It has gone through four historical stages: primitive society, slave society, feudal society and modern times. In the primitive society, due to the low productivity, it went through an extremely slow evolution. The ancestors of Ningbo began with the Hemudu stilt architecture. After a long exploration, they gradually mastered the technology of building ground houses, and invented the original wooden frame architecture, which could meet the basic survival needs. In the slave society, the use of a large number of labor and tools led to a great development of construction activities, with the emergence of such architectural types as Gouzhang city and palaces, and the initial formation of buildings of rammed earth walls and wooden frame. In feudal society, after long-term development and evolution, architecture in Ningbo gradually formed its own style and a mature and unique system. In the first year of Changqing reign of the Tang dynasty (821), Ningbo City was officially built, and many water conservancy facilities and Buddhist buildings were constructed, of which Tuoshan Weir and Pagoda of Tianning Temple are the best examples. The promulgation of *Building Standards* of the Song dynasty promoted the highly standardized development of the construction industry, ushering in the prime time in the history of ancient architecture. The Great Hall of Baoguo Temple in the Northern Song dynasty is known as a physical example and hence a living fossil of the book *Building Standards*. The Hengsheng Stone Arch in present-day Yinzhou District of the Southern Song dynasty is the earliest extant stone arch with an imitation of the wooden frame in China. In the Yuan dynasty, Ningbo, as an important port city in Eastern Zhejiang and at the south estuary of the Grand Canal, witnessed extremely busy shipping traffic. The Site of Yongfeng Warehouse is the site of a large government-owned storage organization in the Yuan dynasty. It is the first site of the Yuan dynasty discovered in China. In 2003, it was rated as one of the top ten national new archaeological discoveries in China. In the Ming dynasty, the scholarly culture was prevalent in Ningbo. Fortunately, the library buildings during that period remained throughout the history. Tianyi Pavilion was the earliest existing private library in China. Tianhou Palace, or Qing'an Guild Hall, of the Qing dynasty was the only building blending a palace and a guild hall among the Major Historical and Cultural Sites Protected at the National Level.

In modern times, the collision of Chinese and Western civilizations brought diverse characteristics to Ningbo's historical buildings. Jiangbei Catholic Church is known as the top church in Zhejiang Province for its Gothic style. Among the modern buildings along the North Riverbank, Ningbo Post Office, the Former Residence of Xie's family, the Former Sites of British

Consulate and Zhejiang Customs are all colonial buildings brought by the British expansion. The latter two were also the earliest Western-style buildings in Zhejiang Province. Qiliyu Lighthouse in Ningbo, one of the lighthouses along the coast of Eastern Zhejiang, is one of the earliest lighthouses built in China and even in the Far East. The Former Residence of Yu's family in Longshan is the largest among the former residences of celebrities of the Ningbo Commercial Group, which can best display the combination of Chinese and Western styles. As the largest and most novel steel bridge in China in the 1930s, Lingqiao Bridge had great contribution to the history of modern bridges in China and in the world as well.

## 2. Scientific Value: Showcasing Ningbo's Innovation in First-Class Building Technologies

Ningbo has a full range of historical buildings. Those in ancient times include urban public buildings, water conservancy facilities, bridges, religious buildings, official residences, folk houses, and gardens in ancient times, and those in modern times include official buildings such as banks and consulates, commercial service buildings, cultural and educational buildings, medical buildings, and the former residences of celebrities of the Ningbo Commercial Group. The construction of all kinds of historical buildings in Ningbo, especially the scientific design and technologies applied, can be rated as world-class.

The stilt buildings unearthed from the Hemudu Site are the earliest wooden structures known in the world to have used the technology of tenon and mortise joinery. At that time, with relatively low productivity, it was remarkable to use such raw materials as wood and stone to build large-scale buildings, which shows the wisdom, creativity and technical level of the ancestors. Tuoshan Weir in the Tang dynasty is one of the four major water conservancy projects in ancient China. Its technology represents a breakthrough for the ancient water conservancy projects in China, more than 200 years earlier than the emergence of similar technologies in other countries. Pagoda of

Tianning Temple in the Tang dynasty is the oldest square-shaped multi-eaved brick pagoda in Jiangnan. As the only example for studying the brick pagoda with twin pagodas in Jiangnan, its corbelled cornices reveal the leading level of brick-making and brickwork of Mingzhou at that time.

The Great Hall of Baoguo Temple is the only well-preserved example of architecture in the Northern Song dynasty in Jiangnan. Many practices and regulations of the Great Hall have become physical examples of *Building Standards*, and some are even the only examples. According to the research of Thomas Young, a British scientist at the end of the 18th century and the beginning of the 19th century, the 3:2 heigh-width ratio of the cross-section of material used for bracket sets in the Great Hall reflects the highest output rate and has the best load-bearing effect. The load-bearing components adopted by Chinese craftsmen are hundreds of years ahead of the experimental data of Thomas Young. Besides, as the official standard of architecture in the Northern Song dynasty, it has already become a standard system that can be described as the most scientific structural modulus.

Up to now, the earliest memorial arches found in Ningbo are Miaogouhou Stone Arch and Hengsheng Stone Arch in Yinzhou District. They are important examples of the transition period from the wooden arch to the stone arch in China. They have a relatively faithful imitation of the wooden structure. Both the roof structure and the details of bracket sets are deliberately treated to imitate wooden structures, which is very different from the stone arches built in the Ming and Qing dynasties. Moreover, there is neither a column base nor a stone flagpole pedestal, which shows obvious characteristics of wooden arches and suggests that these arches are the special structural form characteristic of the in the transition period. They were the first stone arches of the Song dynasty found in Eastern Zhejiang, and their stone construction technology is rare and precious across the country.

Guangji Bridge is the only remaining lounge bridge of the Yuan dynasty. Each of its five bridge piers are made of six bar-shaped stones in parallel, making four arches. Each pier has mortise and tenon joints on both top and bottom and is treated with *cejiao* (approximately 1% slight inward incline of the column on the top). The columns are fixed in the lower part with a whole block of base stone, and in the upper part with *suoshi* (stone lintel beams) to support the wooden beams atop. Its stone craftsmanship can be rated as first-class.

The construction technology of Yongfeng Warehouse in the Yuan dynasty is even more special. The wall foundation of its site is 56 meters long and 16.7 meters wide, covering an area of 940 square meters. The walls around it are specially constructed in the way that square stone blocks with holes in the center are closely arranged at the bottom of the wall, which forms a rectangular building foundation. This type of ancient building structure had never been found before in China. Besides, such a large-scale single building is unique in the archaeological discoveries after the Tang dynasty in China.

In modern times, with the introduction of Western architectural technology, a large number of buildings with mixed Chinese and Western styles have been built. The Gothic architecture of Jiangbei Catholic Church adopts the traditional technology of the column-beam-and-strut framework. Some components used by the former sites of the British Consulate and Zhejiang Customs adopted the early technology of reinforced concrete. Ningbo Drum Tower is the only example in China showing a blend of Chinese and Western styles by integrating the technologies of reinforced concrete and traditional palace-style buildings. Lingqiao Bridges is the largest single-arch steel bridge and the bridge with the most ingenious style in China in the 1930s, which reflects the construction level of steel structure at that time.

### 3. Artistic Value: Shaping Distinct Local Charm Through a Unique Architectural Style

The historical buildings in Ningbo embody many aesthetic principles, with diverse architectural styles, unique forms and distinctive cultural features. Take Pagoda of Tianning Temple for example, above the main body of it are multiple layers of corbelled cornices, which make relatively gentle angles with the pagoda's body compared with the lowest eaves from the ground. The *lingjiao yazi* (chevron-corbelled cornice on brick pagoda) bricks between the layers of projected moldings, the prominent entasis shown in the middle part of the pagoda and gentle tapering at the top, all give the pagoda a towering and beautiful contour. Both techniques of tapering and diminishing treatment are the unique visual artistic treatments to achieve a graceful composition. On the Miaogouhou Stone Arch, the roof is supported by multiple tiers of bracket sets, with upturned eaves and protruding corners. The bracket set in the corner used the *yuanyang jiaojing gong* (the bracket arm that looks like two mandarin ducks crossing their necks in love). It also has a pair of *chiwei* (owl's-tail-shaped ornament at the end of a roof ridge). The arch shows the best-preserved stone carvings with the highest architecturally artistic value among ancient architecture along Dongqian Lake. The art of Three Carvings (wood carving, stone carving and brick carving) of Qing'an Guild Hall in the Qing dynasty are unique in Eastern Zhejiang, especially the finely carved dragon-phoenix stone columns. Lin's Residence built in the Tongzhi reign of the Qing dynasty is the most beautifully carved building among the official residences. Ninghai has the largest number of ancient stages and the most exquisite ones. Besides, they have the architectural style of the triple caisson ceilings (*lit.* algae wells) that is rare in China.

The exquisite traditional carving and the beautiful Western decorations inside the Former Residence of Yu's family in Longshan can be regarded as superb.

The calligraphy in the historical buildings of Ningbo also has a quite high artistic value. For example, the Jukuili Memorial Arch has three large forceful Chinese characters "聚魁里" in regular script inscribed on the front side. There are still couplets written by ancient scholars on the columns of some memorial arches. For instance, the columns of Jiexiao Stele Pavilion (the pavilion with a stele in honor of someone's chastity and filial piety) in Dalu Village, Longguan Township, Yinzhou District, are engraved with the couplets that "She is as flawless as ice and as upright as a well, so the imperial praise is engraved into her arch". Since the Ming and Qing dynasties, most of the bridges in Ningbo have been recorded in stele inscriptions. Scholars and officials of all dynasties have left behind countless *shi* (poems), *ci* (a form of classical Chinese poetry composed to certain tunes in fixed numbers of lines and words), *lian* (couplets) and *fu* (an intricate literary form combining elements of poetry and prose) for the ancient bridges, which have become precious heritage of the bridge culture. On Baiyun Bridge in Luting, Yuyao there are couplets on the outer walls beside both sides of the arch. The couplets on the west read, "Located on the boundary between Yin County and Yuyao County, the bridge sees people celebrating the prosperity there; linking families of Gong and Zheng, the village enjoys thousands of years of peace." The couplets on the east read, "The rainbow of bridge crosses the river that links the north and the south; the moon lingers over the village that decorates the boundary between Yin County and Yuyao County." The bridge couplets not only indicate the location of the bridge, but also express people's wish to live a peaceful and happy life.

The time-honored stages are not only a vivid history of opera, but also a collection of column couplets. Generally, there are couplets written by ancient literati on the two columns in the front of a stage. For example, the couplets of the street stage pavilion at Huanggulin, Yinzhou District, read, "In the transportation hub the opera entertains the passers-by; beside the great port the melodies flow into the water nearby." The couplets on the stage of Temple of Guanyu at the Moon Lake read: "People are watching the opera at the Moon Lake like in the jade pot; stories are played on the stage like the history itself." The thought-provoking contents of the couplets on the ancient stages are either indicative of the environment, or alluding to the present via the past anecdotes.

# Part I

# Primitive Architecture in Ningbo

# Chapter One

# Stilt Architecture: Prototype of Primitive Dwellings

It is said in Chapter "Duke Wen of Teng II" in *Mencius*, "Those who lived in low-lying places made nests, while those who lived on higher ground made caves." Nests were the place where primitive people settled themselves in the humid low-lying areas teemed with insects and snakes.

The primitive architecture in Ningbo developed extremely slowly. Over the long history, starting with the nest-dwelling, the primitive ancestors gradually gained mastery of the technique of building ground houses. They invented the primitive wood buildings, which met the demands for the basic habitation and public activities. It served as the preliminary stage of ancient architecture in Ningbo and laid foundations for its later formation.

Through the archaeological excavation in recent years, some historic sites of ancient Ningbo people's activities have been discovered, indicating that in Ningbo the area of human activities had kept expanding from the Neolithic Age to the Qin dynasty. Even today, many original prototypes of human architecture left by the Ningbo ancestors continue to inspire the modern architecture designing in many respects.

# I. The Excavation and Discoveries of Several Sites of Stilt Buildings

## 1. The Hemudu Site

Large numbers of stilt building sites have been discovered at the Hemudu Site, Yuyao, a county of Ningbo. In particular, at the bottom of the fourth cultural layer, the biggest expansion and the largest number of them were densely and splendidly distributed. Based on the arrangement and orientation of the stilts, architectural experts found that the cultural layer has at least six buildings, one of which over 23 meters long, 6.4 meters deep, with a 1.3-meter-wide veranda under the eave. The long house might have been divided into several small rooms to accommodate a big household. Mainly components of wooden stilts, floor, pillars, beams, and tie-beams were excavated. Hundreds of these components have tenons and mortises, which shows that the mortise-and-tenon joinery has been often adopted in house construction. The buildings at the Hemudu Site are built on the wooden stilts, big and small, on which the big and small beams are put, on which the floor is then laid. In this way the base over the ground is done. Upon the base, columns are erected; upon the columns, beams are installed; and upon the beams, the gable roof is constructed. When the roof truss is completed, it is covered with reed mats or bark for protection.

In these communities, residential area, burial area and pottery-making sites are clearly divided and properly arranged. The floor plan of the houses are round, square or in the shape of Chinese character " 吕 " according to their different functions. The long-house-style buildings built on stilts and with verandas are fit for the humid geographic environment in Southern China. That is why they continued to be used by the later generations. Nowadays, the houses of the same style can still be found in the rural areas in the southeast of China and in Southeast Asia.

## 2. The Tianluoshan Site

The Tianluoshan Site is located close to the Hemudu Site. As a prehistorical village site near the water and mountain that has the best protected ground environment and relatively complete underground remains, it provides a valuable perspective for the study of Hemudu Culture, with multiple layers of the remains of stilt buildings unearthed and the well-organized village layout re-discovered.

In the site, many layers of the remains of stilt buildings have been unearthed, mainly in the form of stilt pits, which demonstrates the features and the level of development of basic construction technology in this stage, as characterized by pits digging, boarding and pillar erecting. The range and size of the architectural remains shows the ancestors' ability to dig deep pits and build houses by utilizing the relationship between gravity and bearing force. Their technology was the most advanced among the Hemudu stilt architecture culture. The site plays a significant role in the research of the technique of wood-frame architecture and that of the development and evolution of ecological environment.

## 3. The Fujiashan Site in Cicheng

The Fujiashan Site is yet another site of a primitive community of the early Hemudu Culture. The foundation of the wood-frame buildings discovered at the site include mostly stilt timbers, wood boards and components with mortises and tenons. It appears that the technique of making these components surpasses that at the Hemudu Site.

The site is built on a wood base, facing east and with Fujiashan Hill at the back. The foundation has a face width of more than 30 meters, which further stretches at the southern and northern ends in the cultural layer, and a depth of over 16 meters with seven to eight rows of wooden stilts at regular intervals. The remains are mostly wooden stilts, wooden boards and components with mortises and tenons, with wooden boards scattered among the wooden stilts.

### 4. The Gouzhang City Site—The Earliest City in Ningbo

Ningbo belonged to the State of Yue in the dynasties of Xia, Shang, and Zhou. The Gouzhang City Site is situated around present-day Wangjiaba Village in Cicheng Town, which had thrived for over 800 years through the dynasties of Zhou, Qin, Han and Jin.

The relics of a timber-frame stilt house were uncovered at the fourth cultural layer, or at the bottom of Trench No. 2 at the site. The building is supported by the crisscrossing stilts piled at the bottom. On the stilts, a layer of board is laid for human activities and enclosed with horizontal timbers, which are reinforced by surrounding posts. As can be seen from the collapsed upper part of the building, there used to be a layer of thatch covered by flat tiles and barrel tiles on the roof. The whole building is rigorously structured and elaborately crafted, which is consistent with the style of Hemudu architecture but possesses unique architectural features typical of Jiangnan water towns. Judging from the relationship between cultural layers, the building was used from the Spring and Autumn period to the Warring States period.

## II. Relatively Advanced Architectural Tools for Carpentry

Hemudu people had an outstanding carpentry. In addition to the wooden spade-like farm tool, small shovel, pestle, spear, oar, mallet, spinning wheel, and knife, wooden handles were also discovered at the Hemudu Site, which were used to fix tools like the bone spade, the stone axe and the stone adze. Among other things, there are also handles in the shape of a try square made of forked tree branches and antlers. Right below the fork head, a tenon is made by cutting and whittling for the binding and fastening of the stone ax on the left and the stone adze on the front. Mortise and tenon joints have been found on many wooden components of buildings unearthed from the Hemudu Site, especially the matchboard with dovetail joints and dowel pin holes, which marks the extraordinary achievements in architectural carpentry in that era.

The stone tools discovered at the Fujiashan Site are mainly production tools, such as stone axes, stone adzes, stone chisels, stone knives, millstones and the like. Bone tools are among the main production tools as well, including arrowheads, spades, daggers, knives, awls and needles all made of animals' limb bones, shoulder bones, ribs or horns, which are the main tools for building the primitive dwellings.

The wood relics unearthed from the Cihu Site, Cicheng, are most rare. The wooden drilling bit (inlaid with the drill made of bone or tooth), as the first-discovered one in history, filled the gap in the research in the development of wood production tools. The wooden winged arrowhead with long cutting edges is speculated as the embryo of the bronze winged arrowhead with short cutting edges, since they resemble each other a lot. And the yoke-shaped tool is probably a type of towing tool. The wood production tools mentioned above play a significant role in the construction of primitive buildings.

# III. Lacquer, Reed Mat, Brick and Tile Widely Used in Construction

More than twenty pieces of lacquerware have been uncovered at the Hemudu Site. In the early stage, they were made by coating natural lacquer over the wooden ware, and later, by mixing red mineral substance into natural lacquer to make colors brighter. The wood-cored lacquer bowl unearthed from the third cultural layer is a typical work, with a coat of vermilion lacquer on the exterior. The painting was found in infrared spectroscopy to have a similar spectrogram with that of the coat of paint unearthed from Mawangdui Han Tomb. Through pyrolysis test, the coating material on the wooden bowl was verified to be raw lacquer. The discovery of vermilion wooden bowls indicates that our ancestors in the Neolithic Age have already recognized the functions of lacquer and had the knowledge of mixing the colors, which has paved the way for the application of lacquer in their construction.

In the tribal groups at Lujiaqiao Site, Siming Mountain in Yinzhou, the Lujiaqiao primitive people have left behind the age of slash-and-burn cultivation, and formed a dwelling village to settle in by the river at the foot of the hills, which is characterized by the use of reed mats for the wall for protection against wind and rain.

In the fourth layer of Trench No. 2 at the Gouzhang City Site, some construction components such as flat tiles, barrel tiles and bricks were discovered. They are so finely made and of such high quality that they could not be used for regular houses. The unearthed eaves-end tiles mainly have the decorative pattern of human face, which is similar to those of the Six dynasties discovered from the then capital Jiankang (nowadays Nanjing). They also belong to the same period in the history. It proves that eaves-end tiles were then used as construction materials in this region.

# IV. Mortise and Tenon Joinery 2,000 Years Further Back in History

The huge stilt buildings excavated at the Hemudu Site are far more complex than the semi-crypt house in the Yellow River Basin in the same age, because the former required a large number of timbers to be calculated, classified, and processed, and an on-site commander was needed on construction. It suggests that Hemudu people were relatively intelligent. The stilt building, which prevents damp air and wild beasts, is the origin of timber structure buildings in Southern China. In particular, the use of mortise and tenon joints, which dated the history of mortise and tenon technology 2,000 years further back, was recognized as a wonder of 7,000 years ago.

Besides the building components with mortise and tenon joints excavated at Fujiashan Site, wooden boards with dowel pins were also discovered. What is even rarer, three trough plates with a circular-arc groove on both sides were found, one end with two square tenons on both sides and the other end flush. It was the first time for such components to be unearthed. The building technique was rather advanced during that age.

Part II

# Ancient Architecture in Ningbo

# Chapter Two

# Xiantong Pagoda: A Testament to Tang Dynasty Prosperity

On West Zhongshan Road stands an ancient pagoda with far less height than Tianfeng Pagoda. It is Pagoda of Tianning Temple, commonly known as Xiantong Pagoda. The ordinary-looking pagoda has an extraordinary history. Constructed in the Xiantong reign of the Tang dynasty (860–873), it is the best-preserved and the only multi-eaved brick pagoda of the Tang dynasty in Jiangnan, and an early extant example of the form of twin towers in front of the temple. Therefore, it is of great value to the study of the history of Chinese architecture, and the history of religious architecture in particular. As a landmark architecture of Ningbo, a famous historical and cultural city, the pagoda and its relic site are among the sixth batch of the Major Historical and Cultural Sites Protected at the National Level.

# I. The History of Xiantong Pagoda

Named after Tianning Temple, Pagoda of Tianning Temple (the west one of the original twin pagodas) is about 210 meters away from the south gate of Zicheng City (inner city) of Mingzhou City (where the present-day Drum Tower stands). Due to the inscription on the pagoda about brick production in the Xiantong reign, it became known as Xiantong Pagoda.

According to the historical documents, when Tianning Temple was established in the fifth year of the Dazhong reign of the Tang dynasty (851), it was first named Guoning Temple. In the Song dynasty, it was renamed Chongning Wanshou Temple in the second year of the Chongning reign (1103), further renamed Tianning Wanshou Temple in the second year of the Zhenghe reign (1112), and then Tianning Bao'en Temple in the seventh year of the Shaoxing reign (1137). It was destroyed and rebuilt repeatedly many times during the Yuan, Ming and Qing dynasties. Along the way, it has been expanded in size. At its peak, the temple had its main gate, main hall, bell tower, drum tower, Qianfo Pavilion, Arhat Hall, Abbot's chamber, iron tower, brick pagoda, meditation abode, and so on. With a farmland of 2,159 *mu* and a woodland of 260 *mu*, it is among the larger temples with the longest history in Ningbo. Initially built in the Xiantong reign of the Tang dynasty (873), the twin pagodas stand symmetrically on the left and the right of the axis in front of the temple. Xiantong Pagoda is the west one.

In 1995, when Zhongshan Road was reconstructed, the archaeologists made an excavation of the Tianning Temple Site and had it preserved and repaired. The work was completed at the end of the year. Since then, it became one of the rare cultural landscapes standing along the east-west main road in Ningbo City.

# II. Reasons Why Only the West Pagoda Remained

Few documents about Xiantong Pagoda can be found. On the bricks in the body of pagoda is the inscription "Made in the fourth year of the Xiantong reign period during the Tang dynasty" in Chinese regular script. According to *Siming Topics* of the Qing dynasty, "Xiantong Brick Pagodas stand on the left and right in front of Tianning Temple." The above-mentioned inscription was also documented in *The General Annals of Yin County* compiled in the period of the Republic of China (1912–1949). It also records that "In June in the third year of the Guangxu reign during the Qing dynasty, the right pagoda fell apart, bricks scattered among folk people. The left one still remains there". The remaining left one is the well-preserved west pagoda.

Then why does the west pagoda still stand while the east pagoda has fallen apart? The archaeologists pointed to the following two possible reasons after through consideration.

Firstly, it is recorded in the documents that Zicheng City of Mingzhou was surrounded by a moat, which was closer to the east pagoda. So possibly, the base of the East pagoda slid toward the moat and sank, leading to its tilt and collapse.

Secondly, as documented in *Siming Topics* of the Qing dynasty, on March 12, 1684, the twenty-third year of the Kangxi reign of the Qing dynasty, a fire in the residential area engulfing the street destroyed the main gate of the temple, the right Sutra-Pitaka, and then the left Sutra-Pitaka. It also severely scorched the east pagoda. Bricks were burned into pieces. The yellow slurry, which served as the adhesive of bricks, were dehydrated, so the adhesion between bricks were weakened. Also, the overhanging wooden beams inside the pagoda were all carbonized, greatly weakening the supporting force from the inside. As a result, the east pagoda collapsed as soon as it slightly tilted to one side.

This was proved by archaeological excavations in 1995. The archaeologists discovered that the fallen east pagoda had a typical Sumeru base of the Tang dynasty on a square plan in the shape of Chinese character " 回 ". Measured from the four angles of the brick base platform, it was found that the pagoda tilted in a twisting way from the base. Based on its height of 12 meters, the pagoda was found to lean over 36 degrees southwest. The central axis tilted more than one degree southwest and the top center of the pagoda had shifted approximately 23.7 centimeters southwest. The bricks of the pagoda were of the same size and quality as the foundation bricks, both black ones, with very few bricks with red in the center. Despite the good quality, after thousands of years of heavy pressure, natural erosion, and deformation caused by changes in humidity, most of the bricks have been crumbling and embrittled, forming numerous cracks.

Within the inner wall of the base, especially at the four corners, supporting walls constructed from numerous bricks of the Han and Jin dynasties were discovered. On the uppermost and the middle layer of the Sumeru base remained some lime. Therefore, it is confirmed that the east pagoda had been reinforced during the Ming and Qing dynasties. At the northeast corner of the underground relics, the archaeologists found a 13-layer fallen brick structure about 60 centimeters square, so they reckoned that the east pagoda had collapsed partially rather than all at once.

# III. Marks of the Flourishing Tang Dynasty on Xiantong Pagoda

Pagoda of Tianning Temple has distinct features of the Tang dynasty architecture. It can be seen from the structure and exterior appearance that the pagoda was constructed during the late Tang dynasty and shows features of brick pagodas of the Tang dynasty.

Pagoda of Tianning Temple Pagoda of Tianning Temple Pagoda possesses the basic characteristics of multi-eaved twin pagodas of the Tang dynasty. The practice prevailed in the Tang dynasty that twin pagodas were built in front of the main hall, either side by side or one behind the other along the central axis. Pagoda of Tianning Temple has a square plan and five courses of eaves. Above the main body are eaves made of corbelled bricks, which project relatively gently from the pagoda. The multi-eaved pagoda of the Tang dynasty is characterized by the *lingjiao yazi* between the layers of projected moldings, the prominent entasis shown in the middle part of the pagoda and the gentle tapering at the top, which give the pagoda a tall and straight appearance.

The base of Tianning Temple was simple in structurem, consisting of the rammed earth layer and the brick foundation layer, above which is the Sumeru base as low as 40 centimeters made of plain bricks. Analysis suggests that earlier pagodas have lower foundation bases about tens of centimeters high. Xuanzang Pagoda in Xi'an is a good case in point. It does not have a visible base, which misleads people into thinking that the pagoda was built over no base at all. By the late Tang dynasty, the base had evolved into the simple Sumeru base, serving both as a visual support for the main structure and as a safety measure.

On the ground floor, all the four sides have a pointed arched doorway. Square-shaped doorways were popular in the early Tang dynasty, while in the late Tang, influenced by the arched caves in tombs, some doorways are made into the shape of pointed arch, which prevailed in Tang pagodas.

The tapering and diminishing techniques adopted by Pagoda of Tianning Temple are also consistent with other brick pagodas built in the Tang dynasty. The tapering technique includes tapering from big to small and from the middle to both the upward and the downward. The former type of tapering involves diminishing in size gradually course by course from the lower part to the upper part until it becomes a pointed tip at the top. The latter one is characterized by entasis, where the middle part, which is the largest in size, diminishes gradually upward and downward to make a graceful outline. The diminishing technique refers to the gradual decrease in the height of each layer of the pagoda, starting from the first or the second story, with the width also decreasing accordingly. The early Chinese pagodas are known for their diminishing technique employed to treat the height of principal stories, which aims to create a visually artistic effect. Tapering and diminishing techniques are unique compositional methods used in Chinese pagodas to achieve aesthetic beauty.

The brick pagodas of the Tang dynasty usually possess the designs of raised eaves and curved bodies. Raised eaves refer to the design and construction where both ends of the eaves across the facade are raised up to create a curve in the otherwise straight eaves, giving a brisk and lively visual effect. The design was first used in timber constructions and later faithfully imitated by brick and stone constructions in the brick eaves. In the design of curved bodies, the straight line of the exterior wall is replaced by the curved outline. Since Songyue Temple Pagoda of the Northern Wei dynasty till the Tang dynasty, curved bodies had been adopted in building the brick pagodas. The view of the pagoda body, the waist eaves, and the top of Pagoda of Tianning Temple shows concave in the middle and a little rise in height on both ends, typical of Tang dynasty pagodas. It is because the design of raised eaves is used for each layer of the eaves and the curved bodies for the pagoda body.

The bricks used by Pagoda of Tianning Temple are of many different sizes, mainly 28–32 centimeters in length, 11–15 centimeters in width, and 3–3.5 centimeters in height, which are similar to those of the same period in the Tang dynasty. In the brick pagodas of the Tang dynasty, bricks of the exterior and interior walls are stacked horizontally in parallel, while the bricks inside the walls are piled up in a comparatively messy way. The bricks are stuck together with damp earth, which is a strongly adhesive substance. By contrast, the mortar applied in the brick pagodas of the Song dynasty is a mixture of yellow clay and white lime. The bricks of Pagoda of Tianning Temple are arranged in the same way as those of the Tang dynasty and only the yellow mortar is applied in laying the bricks without using any white lime.

There is hardly any ornament on the walls of Pagoda of Tianning Temple. The Sumeru base is also plainly designed. Every side of every story has a small Buddhist niche. The corbelled cornices are adopted for the waist eaves, without any decorations or structures such as lotus-blossom columns, *queti*, bracket sets, and carved boards, which are common for the pagodas imitating a timber structure. Besides, the Sumeru base was built with horizontally stacked bricks, and the upper and lower *xiao* (the layer between the uppermost/the lowest and the middle layer ) were also made of bricks. On the west wall of the base of the east pagoda, six bricks with mirrored inscriptions of "Made in the third year of the Xiantong reign" in relief were found, which further confirmed the time when the twin pagodas were established. Between the two pagodas, the relics of a Tang-style road paved with obliquely laid bricks were also discovered, extending northward to the main gate. Therefore, the mysteries were solved about the positional relationship between the temple and the pagoda, the base design, and the road of the Tang dynasty.

While exploring the damages of the west pagoda, archaeologists discovered at the south of its third story some bricks inscribed with Chinese characters "the second year of the Xiantong reign", "the third year of the Xiantong reign", as well as patterns of Chinese character " 米 " (rice). It makes a supplement to the inadequate ancient literature and documents, and a further confirmation that the west pagoda was built in the Xiantong reign during the Tang dynasty. Therefore, it deserves the name Xiantong Pagoda.

# Chapter Three

# A Millennium-old Pagoda: The Epic of Woodwork and Masonry

The pagoda is a common type of ancient architecture. As the old saying goes, "To rescue one person from death is better than to build a seven-storied *futu* for the Buddha." The *futu* in question, namely stupa, refers to Buddhist pagoda. The ancient pagodas all over China are known as outstanding high-rise buildings.

Pagoda means burial mound in ancient India. The Sanskritic name was transliterated into Chinese as *fotu* (Buddha's Figure) and *futu*. The Chinese character "塔" (pagoda) is a figurative name given by the ancient Chinese to this type of buildings introduced from India, which first emerged in the book *The Garden of Words* written by Ge Hong in the Jin dynasty.

According to *The General Annals of Yin County*, in the second year of the Chiwu reign of State of Wu during the Three Kingdoms period (239), Kan Ze from Gouzhang (present-day Ningbo), Grand Mentor of the crown prince of the Wu, donated his residence to the service of Buddha and had it consecrated as a temple, building the first temple in the history of Ningbo—Puji Temple, which is located in present-day Cihu Middle School. In "Inscriptions of Rebuilding

Puji Temple" of the Ming dynasty, it is recorded that "...in the center of (Puji Temple) stands a *futu* and in the east of two pillars the statue of Kan Ze". The *futu* mentioned is the first Buddhist recorded pagoda in the history of Ningbo.

The earliest surviving ancient pagoda in Ningbo is Pagoda of Tianning Temple of the Tang dynasty on Zhongshan Road. In the first year of the Changqing reign of the Tang dynasty (821), the municipal government of Mingzhou was moved from Xiaoxi, Yinjiang, to the Three-river Junction, and Zicheng City of Mingzhou City was built. About more than 30 years later, in less than 100 meters from the city of Mingzhou, a lofty temple was built, yellow walls and black tiles covered in greenery, twin pagodas towering in front of the temple gate. This is the famous temple of the Tang dynasty, Tianning Temple. After a field survey of the twin pagodas of Tianning Temple, the famous expert of ancient architecture Luo Zhewen concluded that it is not only the earliest square brick pagoda in Zhejiang Province, but also the earliest national example of the style of twin pagodas in front of the temple. It is a pity that the east pagoda had collapsed in the Guangxu reign of the Qing dynasty, and only the west one survived.

In the poem "Erling Mountain" (*lit.* Mountain of Two Immortals) written in the Ming dynasty, it reads: "Beside the East Lake is Erling Mountain, / like twin pearls falling into water from a dragon. / Surrounding the mountains are white clouds. / Half the sky is the lone pagoda with rain traces about." It depicts the scenery of "Erling at Sunset", one of the "Ten Views of Qianhu Lake". Whenever the twilight rises at the Dongqian lake, Erling Pagoda is bathed in the afterglow of the sunset and reflected in the quiet water, blending into the scenery of surrounding mountains. Erling Pagoda is not only unique in architecture, but also extremely exquisite in carving. It is a hollow square stone pagoda with a height of nine meters and seven stories. Each story has a waist eave, which is concave in the middle and curved upwards at the ends, with round holes at both ends for hanging wind chimes. In the center of the outer walls of each story is a Buddhist niche. There are totally 39 Buddha statues carved in relief, each with a peaceful look. The first story has three Vajras

(Buddha's guardian warriors), which look tough and majestic. Judging from the inscription of "the ... year of the Zhenghe reign" engraved on the wall of the pagoda and the Buddha statues characterized by Song dynasty carvings, Erling Pagoda is a best piece of work of the Zhenghe reign of the Northern Song dynasty.

If you are on the train heading for Ningbo from Hangzhou, you will see Pengshan Pagoda near Cicheng, a brick pagoda of the Ming dynasty. As the landmark at the city border, it has survived more than 400 years of weathering. An ancient Chinese poem describes it as follows, "Across a hundred *li* of flowers in the fog is a mountain outside the city walls. / With a young lad today I climbed the high mountain. / The stories of pagoda have their shadows reflected in the clouds, / like the fancy fairies in the land between the mountain and the sea." Pengshan Pagoda is so tall and spacious that it can be seen from tens of miles away.

The smallest surviving pagoda in China is the small wooden stupa for Buddhist relics in Ashoka Temple in Ningbo, which is only a few dozen centimeters high, but is a well-known treasure in Buddhism. Legend has it that during the Taishi reign of the Jin dynasty (265–274), a man named Svaha Liu dug out a small black stupa from the ground, "one foot and four inches high and seven inches wide", with a five-disk phase wheel on the *cha* (the top of stupa) and carvings on all the four sides of the stupa, which somewhat resembled a stone. It is said that the stupa is one of the 48,000 stupas made by King Ashoka, and it contains the relics of Shakyamuni. In the second year of the Yuanjia reign of the Liu-Song dynasty in the Six Dynasties period (425) when Ashoka Temple was built, a hall was specially built for the Buddhist relics. With the height of more than 50 meters, double eaves and glazed tile roof, the hall contains a stone stupa, inside which is a wooden stupa inlaid with the seven treasures, and inside the wooden stupa is finally the small stupa for the relics. It suggests that the stupa was highly treasured. The small wooden stupa has Indian-style carvings on it, which is consistent with the Buddhist classics, but it also very much resembles the 84,000 sutra pagodas for the treasure chests constructed by the King Qian Hong of the State of Wu Yue during the Five Dynasties period. So, was it made by the King of Wu Yue during the Five Dynasties period, or was it dug out of the ground during the East and West Jin dynasties, as the legend goes? It remains unknown.

In the Ningbo area, Fenghua District, where six ancient pagodas of different styles are still preserved, has the largest number of ancient pagodas.

Among them is Ruifeng Pagoda located on the top of the Nanshan Mountain in Daqiao of Fenghua. It is a stone pagoda originally built in the Tang dynasty and rebuilt in the Qing dynasty. Next to the pagoda is a stele pavilion. The couplets on its pillar read, "The pagoda soars into the sky like a writing brush; the stele has shined on Nanshan Mountain for thousands of years."

On the top of Yongshan Mountain, about 500 meters west of Jiangkou Town in Fenghua, stands Shoufeng Pagoda, which was once used for military purposes. According to *The Supplementary Annals of Fenghua County*: "Assistant Director of the Left in the Depart of State Affairs, Mr. Tong of the late Tang built a pagoda near the edge of Yongshan Mountain, for the purpose of looking out for the fire beacon and as the tower of *wenfeng* (*lit.* Peak of Literature, which is believed to bring good luck to those taking literary examinations)." The existing pagoda was rebuilt in the twentieth year of the Daoguang reign of the Qing dynasty (1840).

The Stone Pagoda of Dongshan Temple in Cixi City is the most exquisite surviving stone pagoda of the Song dynasty in Eastern Zhejiang, and it has the carvings exactly the same as those of Todaiji Temple Stone Pagoda in Nara, Japan, which is a World Cultural Heritage. A survey shows that only two stone pagodas of Song dynasty still exist in Eastern Zhejiang. The other one is Erling Pagoda in Dongqian Lake, but its carving is far less exquisite than that of the Stone Pagoda of Dongshan Temple.

The Stone Pagoda of Dongshan Temple was originally built with seven stories of carved stones, but now only five stories remain. The pagoda is hexagonal in plan, with exquisite relief carvings of Buddha on each side. It has a total of more than thirty Buddhist statues. It has upturned roof wings and clear-patterned eaves-end tiles. Each story of the pagoda and each of the waist eaves are made of one whole block of stone. The pagoda is simple and elegant in design with high artistic value, and such pagodas are very rare in Zhejiang.

According to the author's survey in recent years, more than 30 ancient pagodas on record survived in Ningbo, which not only have a long history, but also have very different architectural styles, such as being round, square and hexagonal in plan. They often have an odd number of stories, varying from five to thirteen. For example, Pagoda of Tianning Temple has five stories, and Pengshan Pagoda and some ancient pagodas in Fenghua have seven. Seen from the construction materials, there are wooden pagodas, stone pagodas and brick pagodas. Most of the ancient towers in Ningbo are brick pagodas. According to the architectural style, there are pavilion style, multi-eaved style, and the combination of pavilion and multi-eaved styles.

The pavilion pagoda is the oldest, largest and most preserved of all the ancient pagodas in China, and is a unique style of pagoda architecture for the Han Chinese. For example, Ningbo Tianfeng Pagoda was built in the years of *Tian-ce-wan-sui* and *Wan-sui-deng-feng* during Wu Zetian's reign of the Tang dynasty (695–696). The name of the pagoda "Tianfeng" comes from the initial Chinese character of "Tian-ce-wan-sui" and the last one of "Wan-sui-deng-feng". The pagoda is 18 feet high, about 51 meters. It has 14 stories, 7 *mingceng* (*lit.* bright stories, or those that can be recognized from the outside) and 7 *anceng* (*lit.* dark stories, or those that are not marked on the outside), including the underground palace. It is hexagonal in plan, with relatively large spacing between each story. At a glance it looks like a high-rise pavilion. Due to its taller and larger shape, the pagoda is equipped with brick-and-stone or wooden stairs for people to ascend the tower for a distant view. The number of exterior stories of the pagoda is usually consistent with the number of interior stories. Outside the pagoda there are also structures such as doors, windows and columns deliberately imitating wooden structure.

The existing Tianfeng Pagoda is an imitation of typical pavilion pagoda of the Song dynasty unique to Jiangnan in China, exquisite and delicate, simple and solemn. Tianfeng Pagoda was also a navigation mark for river and sea transportation in ancient Mingzhou Port, and therefore an important symbol of the port city. Since the Tang dynasty, due to its prosperity, Mingzhou Port became one of the famous foreign trade ports in China. Foreign envoys, foreign students and merchants made their way to the port from Mingzhou Port and reached Beijing directly by way of Eastern Zhejiang Canal and then the Beijing-Hangzhou Grand Canal. A Frenchman once described in *A Field Survey of China's Export Trade* that the most beautiful city of Ningbo in China had a large number of historical sites, the most striking one being Chifeng Pagoda (i.e., Tianfeng Pagoda), and that on its walls were found the names inscribed by a number of seamen of the French three-sail ship Alkmena, which had visited Ningbo in the previous year. Tianfeng Pagoda is a witness to history and a significant piece of cultural relic of the Maritime Silk Road.

Multi-eaved pagodas are second in number and status to pavilion pagodas among the ancient pagodas in China. Take Pagoda of Tianning Temple in Ningbo for example. It was formed during the evolution from the wooden pavilion pagoda to the brick and stone pagoda. This type is characterized by a high-rise first story, while the upper stories are closely superposed one above another. It is hollow inside and cannot be ascended.

Nowadays, Ningbo ancient pagodas have been effectively protected during the reconstruction of the old city, with dozens of them listed as historical and cultural sites protected at different levels. Since 1995, in order to protect Qishi Pagoda in front of the Qita Temple and Pagoda of Tianning Temple of the Tang dynasty, the municipal government has deliberately allowed the roads to make a turn in front of the pagodas during the reconstruction of Baizhang Road and Zhongshan Road. These bends have become a very special cultural sight in the city streets.

Ningbo's ancient pagodas in various styles present a journey of beauty. The beauty of ancient pagodas showcases the spirit of the times, the power of human creativity, wisdom and faith, and the evolution of each historical stage. It has become an artistic epic of woodwork and masonry.

# Chapter Four

# Great Hall of Baoguo Temple: A Masterpiece of Song Dynasty Architecture

"In the clouds the temple is hidden, staring down on the road of a hundred turns." (Cited from "Touring Baoguo Temple" by Qian Wenjian of the Ming dynasty.)

These two lines vividly depict the beautiful scenery of the secluded Baoguo Temple.

In 1961, Baoguo Temple was announced by the State Council as one of the first batch of Major Historical and Cultural Sites Protected at the National Level, the first one in Ningbo.

As the best-preserved timber structure of the Northern Song dynasty in Jiangnan, the Great Hall of Baoguo Temple has always been the pride of Ningbo people.

Firstly, Baoguo Temple has a long history. According to the literature, in the Eastern Han dynasty, Zhang Qifang, a *Zhongshu Lang* (inner secretarial court gentleman), abandoned his official post and started living here in seclusion, who then donated his residence to the service of Buddha and had it consecrated as a temple, naming it Lingshan Temple, which was later abolished and destroyed. After being rebuilt later, it was then destroyed again in the time of Buddhism persecution during the Huichang reign of the Tang dynasty. In the first year of Guangming reign of the Tang dynasty (880), the local people sent monk Kegong from Guoning Temple in Mingzhou to Chang'an to ask Emperor Xizong of Tang to restore the temple. When Kegong arrived in Chang'an, he gave lectures on Buddhist classics at Hongfu Temple, and the Buddhism were greatly revived. It is said that Emperor Xizong was pleased, thinking that Buddhism was good for the country and Kegong showed meritorious courage to protect the country, so he bestowed the plaque inscribed with " 保国 " (*lit.* protect the country) to the Lingshan Temple. Since then, the name "Lingshan Temple" was changed into "Baoguo Temple". But it was destroyed again later. In the sixth year of the Dazhong Xiangfu reign of the Northern Song dynasty (1013), the Great Hall and other buildings were rebuilt by monk Dexian.

Nowadays, the ancient building complex of Baoguo Temple covers an area of 20,000 square meters, rigorously laid out on three axes. It is so well arranged that it looks both magnificent and elegant. The survived ancient buildings are from many historical periods, including Piaoqi Well of the Han dynasty, Dhanari columns of the Tang dynasty, the Great Hall of the Song dynasty, Yingxun Building of the Ming dynasty, Tianwang Hall of the Four Heavenly Kings, Hall of Guanyin Bodhisattva, Bell Tower, and Drum Tower of the Qing dynasty, and Buddhist Scriptures Pavillion of the period of the Republic of China. It is a miracle for the main hall, which was rebuilt in 1013, to have survived thousands of years of wars and disasters, and even the humid climate and termites typical of Southern China.

Secondly, Baoguo Temple was discovered "by chance". In August 1954, Qi Deyao, then a student of Nanjing Engineering College, and his classmates Dou Xuezhi and Fang Changyuan formed a summer internship team to investigate the folk houses and the ancient architecture of Eastern Zhejiang in Ningbo. Towards the end of the survey, they overheard someone say that there was an age-old "hall without beams" in the mountain north of Hongtang. So they decided to make a trip there in Lingshan Mountain. It was this visit that uncovered a secret that had been forgotten for more than 900 years.

Unfortunately, it was raining heavily that day. Still, supported by enthusiasm, after walking in the rain for half an hour, the three of them finally saw a large temple with black tile roofs up the mountains.

Entering the ancient temple, they carefully examined the structure of the Great Hall. Judging from the bracket sets, caisson ceiling, *gualeng* (*lit.* melon-corrugated) columns and other details which they had never seen, they concluded that this building was unusual. But it was too late to conduct any further survey, do the drawing, or take pictures. As soon as the rain stopped, they took a bus back to Ningbo. The next day when they returned to Nanjing and reported it to their teacher, Professor Liu Dunzhen, a famous architect and member of the Chinese Academy of Sciences, he was so amazed that he made a decision that they return to the temple for a thorough and detailed survey, photography and data collection. Since then, the wonder of Baoguo Temple has been widely known.

Thirdly, when it comes to the history of ancient Chinese architecture, it is impossible not to mention the book *Building Standards*, which was published in the second year of the Chongning reign of the Northern Song dynasty (1103). This officially-published book on architectural design and construction specification is the book with the most complete building technology in ancient China. Whenever talking about this book, we have to mention Wang Anshi, an outstanding politician, thinker and literary scholar of the Northern Song dynasty, known for his famous reform. He was the originator of *Building Standards* and had a historical connection with Ningbo.

In the seventh year of the Qingli reign of the Northern Song dynasty (1047), Wang Anshi came to Yin County as the District Magistrate. In spite of only three years on the position, he did a great job. In a sense, Yin County was a testing ground for his reform. He achieved remarkable results, and it became valuable experience for his further innovation and reform.

More than a hundred years after the founding of the Northern Song dynasty, palaces, government offices, temples, and gardens were built one after another, with luxurious and exquisite designs. The officials responsible for the project, superior and inferior, became corrupt, resulting in great pressure on the treasury to cope with the huge expenses. In order to prevent corruption and theft, there was an urgent need to establish the standards and principles for various designs of buildings, materials, construction quotas and quality targets, so as to clarify the hierarchy of house construction and the art form of architecture, and to specify the strict standard estimates for labor and material.

In the fourth lunar month in 1068, the first year of the Xining reign of the Northern Song dynasty, Wang Anshi arrived in the capital to embark on his reform and established a new system to enrich the country and strengthen the army, in an effort to change the country's poverty and weakness. Wang Anshi's reforms involved the management of the construction industry on a large scale at that time, including the compilation of *Building Standards*, China's earliest technical manual on buildings. Many of the building rules and regulations in the book came from Wang Anshi's working experience in Yin County. However, because the officials involved in the compilation of the book were not competent enough, the book had some flaws in the content and thus failed to be implemented. After the enthronement of Emperor Zhezong of Song (1086), he ordered Li Jie to re-compile the book, which was completed in the third year of the Yuanfu reign (1100), and published in the second year of the Chongning reign (1103).

Fourthly, the construction of the Great Hall of Baoguo Temple is 90 years earlier than the publication of Building Standards. Experts found that the construction of the temple is consistent with the building codes set by *Building Standards*. It suggests that the building technology of Ningbo has been included in *Building Standards*. From another point of view, to this day, many practices, codes and regulations employed in the construction of the Great Hall of Baoguo Temple have become physical examples of the book, and some have even been the only examples.

The Song-style architectural features of the Great Hall are mainly as follows.

Firstly, it has greater depth than width in ground plan.

Baoguo Temple has smaller width than depth in ground plan. This form is extremely rarely found in the surviving timber-frame buildings of the Tang, Song, Liao, Jin, Yuan dynasties, but the two extant timber-frame buildings of the Yuan dynasty in Zhejiang, the main hall of Yanfu Temple in Wuyi, and the main hall of Tianning Temple in Jinhua, have a ground plan the same as or similar to that of the Great Hall of Baoguo Temple. In addition, the main hall of Zhenru Temple of the Yuan dynasty in Shanghai also has a similar ground plan. In the existing timber-frame buildings of the Song and Yuan dynasties, except for the Yuhuang Hall of Yuhuang Temple in Gaoping County, Shanxi, all the buildings having greater depth than width are located in Zhejiang province. They are the earliest buildings among the same form. Therefore, it can be regarded as a common practice in the Song and Yuan dynasties in Jiangsu and Zhejiang.

Secondly, the Song-style bracket set, with a cross-sectional height : width ratio of 3 : 2, is designed for the highest output rate and the best supporting effect.

At the opening of the major carpentry chapter of *Building Standards*, it is recorded that "*cai* is the basic modular unit for all measurements in house building. There are eight different sizes, or grades, of *cai*, which are determined by the type and official rank of the building to be erected". Here "*cai*" is similar to the construction modulus in modern times. The *cai* used for bracket sets of the Great Hall of Baoguo Temple is close to the fifth-grade material. Its *cai* and *qi* (a six-*fen* gap or filler between two *cai*) of the bracket set basically conform to the rules established by *Building Standards*. Besides, its cross-section has a height : width ratio of 3 : 2. According to the research conducted by British scientist Thomas Young in the late 18th and early 19th centuries, such a ratio can ensure the highest output rate and the best supporting effect. The supporting components of this specification used by Chinese craftsmen were hundreds of years ahead of Thomas Young's experimental findings. As the official standard of Northern Song architecture, it has long become the standard rule, which is considered the most scientific structural modulus.

Thirdly, the four-segmented *gualeng* column (melon-shaped column) serves as the earliest example of this type in China, with the smaller-sized timbers spliced together to make the larger-sized component, and the design of *cejiao* (approximately 1% slight inward incline of the column on the top) is obviously adopted for the columns.

The columns of the Great Hall of Baoguo Temple are the components characteristic of that time and that place. The 16 columns in the hall are made of smaller timbers pieced together and then embedded, with their cross-section in the shape of *gualeng*. The number of *gualeng* segments varies depending on the location of the column. In the broad sense, there are two types: The type of eight-segmented *gualeng* on the cross-section is used for the eave column and the inner column; the type of half *gualeng* or quarter *gualeng* are used for the gable column and the back eave column, with the corrugated surface of *gualeng* segments on the outside and the common arc-shaped surface on the inside of the temple. Although this treatment is not recorded in *Building Standards*, it already existed in embryo in the Han dynasty and the Three Kingdoms period. Given the similarities in the shape and use between the *gualeng* columns and the *shuzhu* columns of the Han dynasty, it is likely that they are related to each other. Among the surviving Song dynasty buildings, *gualeng* columns are quite commonly used by buildings in the south, such as Nanping Pagoda in Lin'an, Zhejiang built in the Xining reign of the Northern Song dynasty, Shuinan Pagoda of Nanjian Temple in Fuqing, Fujian Province built in the Xuanhe reign of the Northern Song dynasty, and the stone pagoda inside Feiying Pagoda in Huzhou, Zhejiang built in the Shaoxing reign of the Southern Song dynasty. They all have stone or brick *gualeng* columns. But the rare wooden *gualeng* columns were only found on the Great Hall of Baoguo Temple.

Fourthly, the column bases are roughly the same as those of the buildings in the same period of the Song dynasty.

There are generally three types of column bases — drum-shaped, Sumeru-style and basin-shaped. Among the bases of Sumeru-style, some have carved patterns, and some are plain without any. The basin-shaped column bases of the Great Hall of Baoguo Temple are roughly the same as those in the Sanqing Hall of Xuanmiao Temple in Suzhou, Jiangsu Province (built in 1179) and the main hall of Hualin Temple in Fuzhou, Fujian Province (built in 964), typical of the Song dynasty style.

Three caisson ceilings are wonderfully placed on the front part of the ceiling, under which is the space for Buddhist worship. This structure is unique to the Great Hall of Baoguo Temple.

Ancient Chinese buildings generally have ceilings, and only a few buildings do not. The ceiling serves to block the dust from the eaves, originally called *chengchen* (the ceiling that serves to block the dust from the eaves), and thus to keep it clean below. It later developed into something decorative. As recorded in Ying Shao's book *Customs and Traditions* in the Eastern Han dynasty: "Nowadays the temple hall contains *tianjing*. *Jing* (well) is the representative of the East Well. *Ling* (water caltrop) is an aquatic plant. Given their relation with water, they are believed to avert the threat of fire." The East well is the Well Mansion, one of the Twenty-Eight Mansions, which was believed to control water. They built wells at the highest part of the hall of temples and pavilions, and decorated them with lotus, *ling*, lotus root and algae, thus building the caisson ceiling, in the belief that it can suppress the fire demon. The caisson ceiling was passed down from the Han dynasty. Three nested, hollowed-out caisson ceilings were delicately placed on the ceiling in the front of the Great Hall of Baoguo Temple, making the space for Buddhist worship look magnificent. The earliest caisson ceiling discovered so far seems to be the one in the main hall of Guanyin Pavilion of Dule Temple in the Liao dynasty. Buildings of the Tang dynasty do not contain any caisson ceiling. For example, there is no caisson ceiling in the main hall of the Foguang Temple, only a ceiling or *chengchen*.

*Building Standards* has standardized the treatment of caisson ceiling that the eight larger *yangma* (the major beams) meet together at the top. This form is followed in Baoguo Temple. Moreover, the material used for the caisson ceiling of Baoguo Temple is the *cai* of the seventh grade described in *Building Standards*, which is the only extant example of building caisson ceilings using materials as required by *Building Standards* in the Song, Liao, and Jin dynasties.

Fifthly, some architectural details of the Great Hall of Baoguo Temple resemble those listed in *Building Standards*, some being the only extant examples in China.

There is a component called *chandu chuomu*, or *chuomu* tie-beam, whose shape resembles that of cicada's belly, which can be found nowhere else but in the Great Hall of Baoguo Temple. There are also color paintings of *qizhu babai* (*lit.* seven vermilion and eight white). Color paintings in the Song dynasty includes five major types, namely, *wucai bianzhuang* (blue, green, red, yellow and white all over the architectural components), *nianyu zhuang* (mainly blue and green, resembling the color of jade), *qinglü yunzhuang* (green, blue and the blending of them as the transition), *jielü jiehua zhuang* (red paintings with blue and green margins on architectural components), and the decoration of *tuhong shuashi* (plastering in red, yellow and white). *Qizhu babai* is a type of of *tuhong shuashi*, where white horizontal rectangles are painted in line on the red underpainting. *Qizhu babai* in the Great Hall of Baoguo Temple are the only example of Song-style paintings in Ningbo. In addition, some beam frames and shoulders of the lintels are treated with entasis. This uncommon practice is not found in the early northern buildings, but found in Southern and Northern China in later periods. The treatment is recorded as "entasis on both shoulders" in *Building Standards*.

# Chapter Five

# Ancient Memorial Arches in Ningbo: Precious Living History Books

Some Ningbo people are old enough to know that there used to be lofty stone memorial arches in various forms in many places in Ningbo. "The stone memorial arch at the corner of Sanzhi Street is a historical site, on which tigers and dragons are carved." These lines described the memorial arch in memory of Minister Zhang Zan, which stands in front of the previous Temple of Qimu Jiangjun located at the Moon Lake. Some of the other memorial arches used to stand across a main road, some at the section to a lane, and some in front of the path to a cemetery. Although many of them no longer exist, they can still be found in the names of streets and lanes. A small alley next to Tianfeng Pagoda is named Pailou Lane (*lit.* arch lane).

As Liang Sicheng once said, the city gates, memorial arches, and *pailou* (another type of arch) constitute the unique landscape of the ancient streets of Beijing. The city gate constitutes the oppositive scenery to the main street, where multiple arches and pailou turn the monotonous straight streets into orderly and rich spaces, achieving a similar effect of sculptures, triumphal arches and obelisks in the Western urban streets, which make beautiful embellishments and landmarks in the city.

Until the period of the Republic of China, there still remained a considerable number of stone memorial arches in urban Ningbo, which can be inferred from the long list of records of memorial arches in *The General Annals of Yin County*. Nowadays, however, only a few stone arches have survived in Ningbo City. As the city streets are getting increasingly wider, these ancient "beautiful embellishments and landmarks in the city" are hard to find.

# I. The Origin of Memorial Arches

*Paifang*, the memorial arch, also known as *pailou*, is an iconic open building in ancient Chinese architecture composed of single row or multiple rows of vertical columns and horizontal architraves. In fact, the memorial arch is simpler than *pailou*, since it has no bracket sets or eaves. Traditionally, such buildings without bracket sets or eaves are mostly called *pailou* in Northern China, but they are named *paifang* in Southern China, with or without bracket sets and eaves.

The memorial arch has a very long history. According to *The Rites of Zhou*, the basic neighborhood unit in the Zhou dynasty was known as *lilü*, which was renamed *lifang* in the Tang dynasty. The system of *lifang* (or *lilü*) required the layout of the city to be planned in square grids (in the form of chessboard). Every square of the grid is a neighborhood called *lifang*, which had an equal area of land and was enclosed by walls with a front and a back gate, i.e., *fangmen* (*lit.* gate of the grid). The gates were generally high, with columns and architraves, on which words could be carved. This was the memorial arch in the fledgling stage.

Since the *lifang* system was abolished in the Northern Song dynasty, the gates of *lifang* had become independent gates. And as the actual function of curfew no longer existed, the gate of *lifang* accordingly evolved into a symbolic landmark. From the Song dynasty, through the Yuan, Ming and Qing dynasties, the symbolic and monumental functions of the gates have been constantly enhanced. Also, their structure, construction and decoration have become more and more exquisite and refined. There emerged many types of arches, from single-bay to multi-bay ones and from one-story to multi-story ones, as well as many different combinations of ground plans and materials.

As a witness of society, Chinese arches have their functions and values in politics, economy and culture to different degrees. They also show religious characteristics and reflect the rituals and laws of the feudal society. Moreover, they serve as the sources of valuable historical data and therefore a living history book.

# II. Some Representative Arches in Ningbo

The Confucius Temple in Cicheng, Jiangbei District is the best-preserved one in Eastern Zhejiang, and its gate is what people call Lingxing Gate. It is a special type of arch. According to relevant documents, it was stipulated as early as in the reign of Emperor Gaozu of the Han dynasty that the sacrifice to the heaven should be preceded by the sacrifice to the Lingxing Star. When the altar was built in the sixth year of the Tiansheng reign of the Northern Song dynasty (1028), the external walls of the altar was set up as well as the Lingxing Gate in the shape of an arch. Later, the Lingxing Gate was moved to the Confucius Temple, with the intention of honoring Confucius with the ritual of offering sacrifices to heaven, and because the gate was shaped like a window lattice, the initial character " 灵 " (*ling*, spirit) in the name "Lingxing Gate was altered to a different Chinese character of the same pronunciation " 棂 " (lattice). Therefore the original " 灵星门 " was renamed " 棂星门 ".

The most typical pavilion-style memorial arch is the stone pavilion inscribed with " 钟 郝 遗 徽 ". Built in the first year of the Xianfeng reign of the Qing dynasty (1851), it is located at the north corner of Yuelou'ao Village, Xiangshan County. It is a two-storied all-stone pavilion with a gable-and-hip roof, with a square plan. Facing south, the front top is engraved with Chinese characters of *"zhonghao yihui"* . It has four pillars, on which couplets are engraved. Inside the pavilion stands a stone stele carved with four characters " 坤德永贞 " (*kunde yongzhen*). The stele is 173 centimeters high, 82 centimeters wide, and 18 centimeters thick. The inscriptions read, "Erected by the Grand Coordinator and the Provincial Administration Commission of Zhejiang, Headquarters of Ningbo Prefecture and Xiangshan County". According to *Xiangshan Journal* in the period of the Republic of China, the stone pavilion was dedicated to Huang, the wife of Zhuang Yaoxu.

Jiexiao Stele Pavilion, Yinzhou District, is located to the east of the ancestral hall in Dalu Village, Longguan. Built in the Qing dynasty, it is a one-bay structure, 4.5 meters high, 2.1 meters wide, and 1.26 meters in depth. The pavilion has a gable-and-hip roof, with interlocked barrel tiles, and upturned eaves on roof corners, and the two ends of the roof ridge are decorated with *wenshou* (zoomorphic ornaments). Right under the eaves is a vertical tablet carved with dragons. Over the dragon pattern is the engraved Chinese " 圣旨 " (imperial edict), and at the lower place it reads "Application submitted by the county school in the tenth lunar month of the fourth year of the Daoguang reign; endorsed by all levels of officials in the twelfth lunar month of the fourth year of the Daoguang reign; submitted to the Emperor and awarded on imperial orders by the Ministry of Rites in the twelfth lunar month of the fifth year of the Daoguang reign". Surrounded by black stone railings, the square pillars of the pavilion are engraved with the couplets "She is as flawless as ice and as upright as a well, so the imperial edict in red are engraved into her arch". In the center of the stone pavilion stands an upright stone stele, with the characters " 钦旌 " (granted by the emperor in person) and " 节孝 " (chastity and filial piety) engraved on the front and "This stele is erected in honor of Lady Dong of the

Chen family for her chastity and filial piety" on the back, with the main text about her life and deeds. And the inscription reads, "Written by Nephew Cui... in the fourth lunar month of the twenty-ninth year of the Daoguang reign."

The most typical two-pillar and one-bay stone arch is Enrong Arch in Cicheng. It has a gable-and-hip roof, with a height of 6 meters and a width of 3.5 meters. It has a stone roof with barrel tiles and two fish-shaped *chiwei* on the principal ridges. A plaque was hung at the middle front of the upper architrave, which was engraved with a pattern of double dragons playing with a pearl, over which are the characters " 圣旨 " in intaglio. The lower architrave has a relief carving of two lions dancing with embroidery ball in the middle, and a dragon head on each side carved in relief. The upper architrave on the back also has a plaque with the image of double dragons playing with a pearl, engraved with characters " 恩荣 " (the grace and glory of the emperor) in intaglio. On the middle architrave, horizontally inscribed in intaglio are four characters " 诰封 三代 " (conferment of honorary titles for three generations by imperial mandate). The north end of the architrave has the inscription marking the construction time "On a blessed day in the Qianlong reign", and the south end is inscribed with the intaglio signature "Erected by Xiang Hengsheng, grandson of *rulinlang* (a sixth-grade official), also Associate Administrator".

The arch was built by Xiang Hengsheng in honor of his grandfather Xiang Tengjiao, who was a *jinshi* (a scholar who passed the highest imperial examination) in the eighteenth year of the Shunzhi reign of the Qing dynasty, and served as a *shoubei* (a military commander of a city), *youji* (a general), and so on. With more than thirty years in office, he returned home at an old age, known as *wanjie* (the completion of duty with moral integrity). In recognition of his achievements, Emperor Qianlong conferred on him the title of General Wuji (Courageous Cavalryman), and mandated the establishment of the arch.

The most typical stone arch with four columns, three bays and triple eaves is the one with the inscription " 高风千古 " (immortal noble character) located in Huangqingyan Village, Ditang Town, Yuyao City, whose eaves have been destroyed. It has a full width of 8.7 meters and a height of 6 meters. The upper architrave is engraved with four characters "高风千古", while the lower one is inscribed with "Dedicated to Mr. Yan Ziling, the *zhengshi* (soldier on expedition) in the Han dynasty". The east and west bays are carved with lions rolling embroidery ball, birds and animals, and other patterns in openwork caring. The stone arch is so magnificent that it displays the excellent stone carving craftsmanship in the Ming dynasty and is of great artistic value. In the thirty-second year of the Wanli reign of the Ming dynasty (1604), while restoring the ancestral hall and tomb of Master Yan Ziling, the Surveillance Commission of Zhejiang Province rebuilt the arch specially in honor of him, a senior master of the Han dynasty.

Caihong Arch is also a triple-eaved stone arch with four columns and three bays, located in the west section of North Caihong Road, in former Jiangdong District (now Yinzhou District), Ningbo. It was built by the imperial court of the Qing dynasty in 1818 to honor Wu Minghao's wife surnamed Bao.

Wu's family lived in Jiangdong for generations and had their own family business. Their store, Wu Damao Soy Sauce Store, after several generations of diligently running, has been well known throughout Ningbo when it was passed down to Wu Minggao, who died young and left his wife, Lady Bao, widowed young. She alone raised their son Wu Qiyuan, who was only six months old then. She taught him herself. Later her son became a government official. After Bao's death, Wu's family built "Jiexiao Arch" (the arch to honor chastity and filial piety) in front of the ancestral hall of Wu's family to glorify the family.

Caihong Arch has four columns and three bays. It is majestic and solemn, mainly composed of such major components as columns, architraves, crossing tie-beam (the tie-beam connecting an eaves column and the adjacent interior column), bracket sets, and *queti*. The two middle columns are about 5.7 meters high and the two side ones are 4.25 meters high. It has an octagonal plan. The characters "节孝" are engraved in intaglio in regular script on the plate. Over the architrave are bracket sets. And on the architrave and other constructional components, some patterns of flowers, animals, and stories are delicately carved in high relief and openwork carving. The roofs are decorated with a pair of *chiwei*, on which is a dragon head, and a ball in the middle of the principal ridge. They make a traditional image of "two dragons playing with a pearl". The vertical ridges are also embellished with small dragon heads, imposing and majestic, giving a sense of profoundness and sacredness. This stone arch provides an important physical example for experts and scholars to study the stone architecture and carving art of the Qing dynasty.

Yingzhou Jiewu Arch is located to the south of Liuting Street, alongside the Moon Lake, Haishu District, Ningbo, with three bays, four columns and three eaves. On the architrave, " 瀛洲接武 " is carved. The tall arch was built by the Grand Coordinator Gan Tujie, among others, in the thirty-ninth year of the Wanli reign of the Ming dynasty (1611) in honor of Yao Zhiguang and other *juren* (successful candidates in the imperial examinations at the provincial level).

Lixinwu Stone Arch in Zhenhai District has a relatively rare style in Ningbo, with four sticking-out columns and three bays. It is located in a field 10 meters away from Lixinwu, Guisi Miaogang Village, Luotuo Sub-district. There used to be two tombs, which have been destroyed. In front of the tombs, there are two stone arches, towering over an area of more than 800 square meters. On the stone arch, there are some fine carvings and patterns. According to the experts' field work and survey, these stone arches were built in the Ming dynasty.

So far, the earliest arches found in Ningbo are Miaogouhou Stone Arch and Hengsheng Stone Arch in Yinzhou District, both built between the Southern Song dynasty and the Yuan dynasty. Miaogouhou Stone Arch is located in Hanling Village, Dongqianhu Town, Yinzhou District, and Hengsheng Stone Arch, built in the second year of the Shaoxing reign of the Southern Song dynasty (1132), is located in Hengsheng Village, Wuxiang Town, Yinzhou District. Both arches are on the way leading to the tomb passages which have been destroyed, and the tomb owners cannot be verified. They both are stone arches imitating a timber structure, with two columns, one bay and one story, facing east. On the Miaogouhou Stone Arch, the roof is supported by multiple layers of bracket sets, with upturned eaves and protruding corners. The bracket set in the corner used the *yuanyang jiaojing gong*, namely the bracket arm that looks like two mandarin ducks crossing their necks in love. It also has a pair of *chiwei*, namely the owl's-tail-shaped ornament at the end of a roof ridge. It has one of the best-preserved stone carvings with the highest architecturally artistic value among architecture along Dongqian Lake. The material used for the arch is Meiyuan rock produced in the west of Yinzhou District. With a width of 3.03 meters, Hengsheng Stone Arch has a similar basic structure to that of Miaogouhou Stone Arch. What makes it different from the other are the following: The upper architrave is lower and inserted into the columns; it has no *pupai* tie-beam (a long, horizontal architectural element placed on top of the architrave to support bracket sets); *chagong* (bracket-arm directly inserted into another member such as the column shaft) is used for *huagong* (any bracket arm in the bracketing structure that projects away from the wall); on the architrave a rectangle-shaped groove was made in the style of *qizhu babai*; the rock material is Yi'ao rock from Dongqianhu Town, Yinzhou District.

These two stone arches are important examples of the transition from wooden arches to stone arches in China. It was relatively late to discover them, so much so that the surrounding constructions have been lost, and there are no records about them in the relevant history books. It is a unanimous view in the academic field that the stone arch in China originated from the wooden arch. After having compared and analyzed their architectural features with those Song dynasty arches in *Building Standards*, experts concluded that the construction of them can be traced back to the Southern Song dynasty, due to the faithful imitation of timber structure, whether it is the roof structure or the detailed treatment of bracket sets, which is quite different from those built in the Qing dynasty. The arches have no column bases or stone flagpole pedestal, showing obvious characteristics of a wooden arch, indicating that they are still in transition period from the wooden arch to the stone arch. Many of the treatments of the two arches are basically in line with the standards of *Building Standards* of the Song dynasty. The discovery of these two arches fills the gap of Song dynasty stone arch in Eastern Zhejiang. They are also rare and precious across the country.

# III. Types and Architectural Features of Ancient Arches in Ningbo

The ancient stone arches in Ningbo not only have a long history, but also have a full range of types, but only a few have been preserved. Statistics show that there are only more than 30 well preserved arches on record in Ningbo, with only six and a half in the old city of Ningbo, namely, Zhenjie Arch on Caihong Road, Yingzhou Jiewu Arch, the arch of the Ming dynasty in front of the tomb of Quan Shaowei, Quan Zuwang's ancestor, in the tomb area of Quan Zuwang, the stone arch with the inscription "In Honor of the Revered Mr. Wan, Commander-in-chief in the Ming Dynasty" and a pair of Tu's arches in Baiyun Manor, with the "half" being the remaining bay of Minister Zhang's Arch of the Ming dynasty on the east bank of the Moon Lake. The others are scattered somewhere in cities and counties. These ancient arches fall into the following five major categories.

First, sign arch. Examples are the Arch of Tu Bingyi's Hometown of the Ming dynasty located on the green belt of Jiangbei People's Road, Simingshan Arch in Longguan Township, Yinzhou District, Shimen Arch in Fenghua City, and Paimenshu Village Arch at Xidian Township, Ninghai.

Second, chastity arch. Examples are Bao's Chastity Arch of the Qing dynasty located on North Caihong Road in Jiangdong District, "Zhonghao Yihui" Stone Pavilion in Xiangshan, Shuangjie Arch in Longguan Township, Yinzhou District, and Liu's Chastity Arch of the Ming dynasty at No. 4 of Shangzhi Road, in Cicheng Town.

Third, the merits and virtues arch. For example, Enrong Arch, Shi'en Arch and Dongguan Arch in Cicheng Town, Gaofeng Qiangu Stone Arch and Jianyi Arch in Shijia Village in the south of Yuyao City, and Minister Zhang Arch of the Ming dynasty and Yingzhou Jiewu Arch by the Moon Lake in Ningbo urban area.

The fourth category is the arch on the tomb passage. This category is the most widely distributed and numerous in Ningbo, such as Shi's Arch in Yinzhou district, Miaogouhou Stone Arch, the Tu Yu's Tomb Passage Arch of the Ming dynasty on Renmin Road in Jiangbei District, the tomb passage arch previously at Zuguan Mountain ( 祖关山 ), the Ming dynasty arch in the tomb area of Quan Zuwang, and Ding Jiansi Arch ( 丁建嗣牌坊 ) of Ming dynasty.

The fifth category is special, including the Lingxing Gate in front of the Confucius Temple in Cicheng and the arch-style county school gatehouse in Yinzhou District.

The building materials used for ancient archs in Ningbo are generally stone, brick and wood, among which only those built with stone were preserved till today. Since Ningbo is near the sea, the air is humid and there are often typhoons, the arches built outdoors with wood and bricks are likely to be destroyed due to their poor durability. The stone structure has strong resistance to weather and erosion; its durable appearance does not only highlight people's tribute to the ancestors but also makes it possible to pass it on to the future generations. For example, the "Gaofeng Qiangu" Stone Arch, 8.7 meters wide, which commemorates Master Yan Ziling of the Han dynasty, has a stone beam that is nearly four meters long and weighs about two tons, suggesting Yan's significance as the backbone of the country.

Most of the ancient arches in Ningbo have three bays and four columns or one bay and two columns. Only a few are the type of four-column pavilion-style arch, of which only three are known so far in the city. They are the pavilion-style stone arches inscribed with " 钟郝遗徽 " in Xiangshan, Jiexiao Stele Pavilion in Longguan Township, Yinzhou District, and the Qinjing Jiexiao Stele Pavilion in Fenghua City, which are all very precious. The architectural style employed in the arches is generally based on the budget, scale and width of the street. Examples are Yingzhou Jiewu Arch on Liuting Street and Rainbow Arch on North Caihong Road; the roads are relatively wide, and accordingly they are both of the four-column and three-bay type. Those in less wide streets and lanes, though, are mostly two-column and one-bay type and relatively small in scale, such as Arch of Tu Bingyi's Hometown which was originally located at the entrance of Tujia Lane. Among the Ningbo ancient arches there are eave-structure arches and arches with sticking-out columns. The former arches have similar widths but different eave types, such as single eave, double eaves and three eaves. The types of five eaves and seven eaves can only be found in the old photographs.

The structure of Ningbo ancient arches is dominated by beams and columns. The eaves do not have deep projections and complicated *douqi*, and are quite moderate. They were generally simple before the Ming dynasty, but gradually became complex after the Qing dynasty. Their ornaments are mainly stone carvings, mostly using the local stone materials in Ningbo, especially "Xiaoxi rock" and "Meiyuan rock" at Yinjiang Bridge and Meiyuan area in Yinzhou, and "Dayin rock" made at Dayin in Yuyao, because they are of good quality and can be chiselled into bar-shaped stones, columns and beams. Like the famous wood carving, the stone carving in Ningbo is also well known for its great diversity of techniques including relief carving, intaglio carving, round carving, openwork carving, and so on, and its lively patterns as delicate and elaborate as the wood carving. Arches in Ningbo adopt openwork carvings mostly in a single layer, which narrates stories, legends or historical events related to the honored persons, which provide contexts for the arch. At the same time, they are also decorated with stone lions, stone drums, and decorative elements of flowers, auspicious animals and religious instruments, producing the thick, heavy, solid and clear-cut look of the whole arches.

The calligraphy on the arches in Ningbo also demonstrates quite a high level of artistic taste. Most of the arches were privately funded, and famous local writers or distinguished men were invited to write the names for the arches. There is a proverb going: "On the bridge one enjoys fried noodles; on the street one appreciates the calligraphy on the arches." The calligraphy on the arches were all written by famed calligraphers. Take Jukuili Arch for example, on the front side are inscribed the large characters "聚魁里" in regular script, and on the upper right side are engraved "Yang Shouchen, the first place in Zhejiang *xiangshi* (the provincial-level imperial examination) in the first year of the Jingtai reign (1450)" and "Yang Shouzhi, the second place in Zhejiang *xiangshi* in the first year of the Chenghua reign (1465)". In the middle of the back side is the same inscription of the name "聚魁里". On the upper part are the inscriptions of "Wang Ping, Officer of Metallurgy Affairs of the Surveillance Commission of Zhejiang Province", "Li Xing, Prefect of Ningbo", and "Han Pu, District Magistrate of Yin County", and on the lower part is "Erected by Yang Maoyuan, Deputy Commissioner of the Surveillance Commission of Huguang Province on the blessed day in the ninth lunar month of the fifth year of the Hongzhi reign (1492)".

On the columns of some arches, the couplets inscribed by ancient literati are still visible, such as the engraved couplets on the columns of Jiexiao Stele Pavilion in Longguan Township, Yinzhou District. Some couplets are engraved to describe the environment, while some to allude to the present by telling the ancient anecdotes.

# Chapter Six
# Ancient Bridges in Ningbo: Stone Rainbows over the Water

"The sand the road to heaven is built; from the rainbow the bridge to the lonely island is made." This is a verse written by Wang Anshi, a well-known minister of the Northern Song dynasty, when he was an official in Ningbo, which describes the ancient bridges that can be found everywhere in the vast land of Eastern Zhejiang.

Ningbo is located in the south of the Yangtze River Delta, and for thousands of years, the industrious and intelligent Ningbo people have built thousands of ingenious and magnificent bridges spanning between the towns and among the mountains, which facilitate the transportation, decorate the rivers and mountains, and have become one of the symbols of ancient civilization.

Bridges are an important part of ancient Chinese architecture. In primitive times, mankind could not cross streams, rivers and valleys without utilizing the rocks, fallen bamboos or trees. According to the ancient people, the bridge was a beam across the river, an overhead structure, so it was also called the "water beam".

The Hemudu people in primitive times were able to use canoes for transport despite the barrier of water. Given that the Hemudu people were able to make the stilt buildings with mortise-and-tenon wooden structure at that time, they should also have had the primitive wooden bridges, bamboo bridges or bridges made of mud and stone.

According to relevant materials, in the third year of the Long'an reign of the Eastern Jin dynasty (399), General Gao Yazhi had a battle against Sun En outside Wusheng Gate, therefore the bridge outside the city gate was named Wusheng Bridge (*lit.* bridge of military victory). This bridge, located in Yuyao City, under which Yaojiang River flows straight into Yongjiang River and then into the sea, is the earliest famed bridge in record in Eastern Zhejiang.

Since the Tang and Song dynasties, the capital city of Mingzhou (now Ningbo) had "three rivers, six artificial rivers and two lakes in the city". The convenient and well-developed water system has fostered people's way of life in Ningbo since ancient times, one that residents "built bridges for the road" and "made boats for the horse". According to "Measurement Map of Water System in Ningbo County" in the Guangxu reign of the Qing dynasty, in the area of Luocheng City (outer city) of Ningbo at that time, that is, in the narrow area surrounded by present-day Changchun Road, Lingqiao Road, Heyi Road, Yongfeng Road, and Wangjing Road, there were nearly 160 bridges. As time passes, many ancient bridges have vanished due to the natural forces and the social development, but some of them have gone survived the passage of time, still telling the history and playing a role. According to the survey made by relevant experts in 1998, nowadays there are more than 500 ancient bridges in Ningbo. Among them, Yinzhou District owns the most, with 130 bridges; Ninghai and Fenghua are the second, with 123 and 79 respectively; as for the rest, 48 bridges are in Yuyao, 33 in Cixi, 31 in Xiangshan, 22 in Jiangbei, 17 in Beilun, 16 in Zhenhai, 14 in Haishu, and 2 in Jiangdong. Undoubtedly, Ningbo is worthy of the title of "kingdom of bridges".

Based on their structures, ancient Chinese bridges can be categorized into girder bridges, arch bridges, pontoon bridges and suspension bridges.

Ningbo's existing ancient bridges are mostly girder bridges and arch bridges. The pontoon bridge can only be seen in historical records. Lingqiao Bridge, which lasted for a thousand years, used to be a pontoon bridge named Dongjin Pontoon Bridge before reconstruction. There is no record for the suspension bridge in history, nor has any example been found.

According to the architectural features, the ancient bridges in Ningbo area include lounge bridges, stone arch bridges, girder bridges and special bridges.

Panchi Bridge is a type of the special bridge, a ritual stone bridge unique to China, which is strictly set in groups of three, with a gentle sloping surface and some steps. It can be an arch bridge or a girder bridge, but it must be built over Panchi Pond in front of the ancient school at county-level or above (also Confucius Temple), connecting the Lingxing Gate and Dacheng Gate. Officials and pupils who had entered the school and worshipped Confucius were qualified to cross the bridge. Therefore, the scholars' being admitted to the school is called *rupan* (*lit.* entering Panchi). The Chinese character "pan" is specially used for the academy and Panchi Pond. The remaining Panchi Bridges in Ningbo are found only in Cicheng, Zhenhai and Ninghai.

Wuji Bridge in Ninghai, as an example of the girder bridge, with 47 arches and a length of 137 meters, is one of the longest multi-arch sea shoal bridges in Ningbo.

Bailiang Bridge (*lit.* bridge of hundreds of beams) in Yinzhou District, the longest lounge bridge in Ningbo, was originally built in the Song dynasty. It has six piers, seven arches and a wooden frame covered with tiles, not only to shelter from the sun and the bad weather, but also to delay its decay, prolong its service, and provide a resting place as well. The site selection and the construction of Bailiang Bridge conform to scientific principles. As a result, hundreds of years later it still remains despite the frequent floods.

The ancient bridges with a long span in Ningbo are complicated in structure. Without the technical drawings, ancient people would construct bridge models while building the bridge. In spite of this, many bridges were built with thoughtful design and rigorous structure, which is in line with scientific principles. Some bridges were delicately built by combining the advantages of several types such as girder bridges, arch bridges, and suspension bridges. For example, the Fuxing Bridge in Fenghua, the largest five-arch stone bridge in Ningbo, with the central arch spanning 15 meters, is a technical example of the ancient stone arch bridge of Ningbo.

Wodu Bridge at Longxishang, Yuanjia'ao Village, Xiaowangmiao Sub-district in Fenghua, was built in 1759 (twenty-fourth year of the Qianlong reign of the Qing dynasty) and rebuilt in the 1920s. Measuring 24 meters long and 6.4 meters wide, it is a single-arch wooden beam lounge bridge, with a "house" built on it. It is currently the only beam-arch bridge found in Ningbo. No pier or rivet is used for the arch, but beams are cleverly arranged and joined into an arch structure, which is unique to China.

Ancient bridges are not only a common yet special architectural entity, but also an embodiment of the local history, customs and culture. The long history, exquisite skills and literature connected to the bridges constitute a colorful and distinctive culture of ancient bridges in Ningbo.

One of the earliest extant ancient bridges in Ningbo is Huide Bridge in Xi'ao. It has no steps, but has a gently sloping arch on each end of the bridge deck, which is the characteristic of Song-dynasty bridges. It is located in Xi'ao Village of Changjie Township in Ninghai County, Ningbo, across the same stream as Citang Bridge and Siqian Bridge. It is the only one of its type in Zhejiang Province.

Huide Bridge was built during the Baoyou reign of the Southern Song dynasty (1253–1258). It is a single-arch stone bridge with a beautiful shape and an exquisite structure. The total length of the bridge is 11.5 meters, the width 4.5 meters, the height 3.6 meters, and the span 7.5 meters.

It is said that in Ninghai where Huanggong Ferry was located, due to the broad port and furious tide, the ferry boats were often overturned, bringing a lot of suffering to the local villagers. Later, in the Baoyou reign of the Southern Song dynasty, many famous people emerged and assumed official posts, who built the bridge to express their wish that their virtues could benefit their township and be passed down to the future generations. Therefore, it was named Huide Bridge (*lit.* bridge of benefit and virtues). Since then, whenever the villagers needed to go to worship their ancestors, they would surely pass this bridge.

For so many years, Huide Bridge has been silently lying over Xi'ao Stream. The bridge name carved in the voussoir was not discovered until recent years. Huide Bridge is beautifully shaped with elaborately treated details. Along the two sides of the bridge are built eight I-shaped stone balustrade panels, separated with a lotus-shaped top of stone baluster, eight in total. After hundreds of years, five of the balusters have been ruined and only remains are left. In the eyes of experts, these lotus-shaped balusters are almost the same with those in front of Shi Miyuan's tomb of the Southern Song dynasty at Dongqian Lake. The four ends of Huide Bridge balustrades are buttressed with four drum-shaped bridge pillows. The bridge deck is covered by curved stone slabs, which have been fractured due to crustal movement. In the middle of the north side of the closing arch are carved the Chinese characters of 惠德桥 (Huide Bridge) in double-line intaglio.

Huide Bridge shows how skillful and delicate the decorations and craftsmanship of the Song dynasty bridges are. It has four *longmen* columns (*lit.* dragon gate column) on both sides. Atop the balusters are four small carved stone lions of the Song dynasty style with small ears, projecting eyes, and nose and lips on the same level, exactly the same as those of Yuling Tomb of the Song dynasty. Therefore, the bridge is also called "Four-lion Bridge" by the villagers. The two lions who face outward have their mouths open, while the other two facing inward have their mouths closed. According to the local villagers, this treatment can spare them from floods and evils. Others say that "having the mouth open" is to instruct people to do more good rather than bad, and to return home in time after their achievement of success and fame instead of forgetting their hometown, while "having the mouth closed" means to abide by the law, follow the rules of the customs and get along with neighbors after returning home. It is also said that the lion heads were carved to commend a virtuous minister in the Southern Song dynasty for his success in flood taming.

After so many years, the pattern of *guijiao suyun* on the side of the bridge has been corroded, but still can be clearly identified as beautiful and elegant. According to the research findings, the *ruyi* (good luck) cloud pattern of *guijiao suyun* is the decoration commonly used for the bottom corner of furniture and architecture of the Song dynasty. The precious Huide Bridge in the original structure is the second stone arch bridge of the Song dynasty discovered in Ningbo, following the initial discovery of the Southern Song dynasty stone arch bridge in front of a tomb at Dongqian Lake.

Guangji Bridge, the only remaining lounge bridge of the Yuan dynasty located in the northwest corner of Nandu Village, Jiangkou Sub-district, Fenghua City, was originally built in the Song dynasty and rebuilt in the twenty-third year of the Zhiyuan reign of the Yuan dynasty (1286), and rebuilt several times again in the Ming and Qing dynasties, but with the piers basically untouched. The bridge is a four-arch lounge bridge of timber and stone structure, stretching across Fenghua River from east to west, 51.68 meters long, and 6.60 meters wide. With 22 columns in the lounge, the bridge looks as light and graceful as a rainbow over the river. On both sides of the approach bridge there are twelve bays. On the bridge there are five-purlin beams, 3.13 meters wide, and two adjacent corridors each measuring 1.8 meters in width, where people can take a rest. The balustrade plaques are made of wooden boards. The deck of the bridge is also paved with wooden boards, while black bricks and bar-shaped stones were used for paving the approach bridge and building the stairs. On the east and west sides of the bridge there are a three-bay house with a flat roof, whose central bay is *lit.* used for passage. In the two side bays on the west stand six steles, including a bridge stele, a Jinyue (rules and regulations) Stele, and a tea stele. The bay on the east side is for firefighting, with firefighting equipment named "yanglong" (*lit.* foreign dragon). The bridge piers are made of bar-shaped stones set in parallel side by side, making four arches. Each of its five bridge piers are made of six bar-shaped stones in parallel, making four arches. Each pier has mortise and tenon joints on both top and bottom and is treated with *cejiao*. The columns are fixed in the lower part with a whole block of base stone, and in the upper part with stone lintel beams to support the wooden beams atop. On a bar-shaped stone is inscribed "Reconstructed at the hours of *yichou* on the twenty-ninth day of the fourth lunar month in the *bingxu* year, the twenty-third year of the Zhiyuan reign" and so on. As the only remaining lounge bridge of the Yuan dynasty in Ningbo, Guangji Bridge enjoys a high historical value.

Bailiang Bridge (*lit.* Hundred-beam Bridge), the lounge bridge with the largest span in Eastern Zhejiang, also one of the Historical and Cultural Sites Protected at the Provincial Level, is located in Huijiang Village, Dongqiao Town, Yinzhou District. It is a wooden beam lounge bridge on stone piers, scientific in construction and strong in structure. It has six piers and seven arches, 77.4 meters long and 8 meters wide. The piers are made of stacked huge rectangular stone, each 7.8 meters wide and 1.7 meters thick. The beams of the bridge are made of big fir timbers with a diameter of 40 to 50 centimeters arranged in a row, each row comprising about 17 or 18 timbers, and therefore totally 124 timbers form the overall skeleton of the bridge, hence the name Bailiang Bridge, or Hundred-beam Bridge. The deck is covered with chestnut planks of five centimeters thick. On both sides of the bridge there are benches for passers-by to take a rest and wooden guardrails for pedestrians' safety. The bridge has 23 bays, each 3.4 meters wide, with 88 round columns and 44 square columns. It has a round ridge roof covered with black tiles, with double gentle slopes, which can better resist the earthquake. On both ends of the bridge are the gate houses with gable-and-hip roofs, with four square stone columns. The beams are exquisitely carved and a plaque painted in black with gold inscriptions is hung up in the middle at each entrance. The plaque at the south is inscribed "The bridge was built in the Song dynasty", next to which is a stone stele engraved with " 光溪與德会记 " (recorded by Yude Society in Guangxi); the plaque at the north reads " 龙眠蕙江 " (a dragon sleeping in Huijiang River), and next to it stands " 奉宪勒石永禁之碑 " (the stele of injunctions by law); all these were built in the Qing dynasty. The original Longwang Hall (*lit.*

dragon king hall), Sanguan Hall, Wenwu Hall, Guanyin Hall, and shrines for God of land and God of wealth on the bridge were restored though funding by local people in 2002. The original Dhanari at the north end of the bridge was destroyed by typhoon in 1956. Part of it is now kept in the District Cultural Administration Office, and the rest of it is in the collection of Tianyi Pavilion Museum.

The most beautifully shaped ancient bridge in Ningbo must be Baiyun Bridge, which is located in Zhongcun village, Luting Township, Yuyao. Originally built in the Zhenguan reign of the Tang dynasty (627–649), it was destroyed and rebuilt several times. The present one was reconstructed in the sixteenth year of the Guangxu reign of the Qing dynasty (1890). Its shape and architectural style are distinctive.

Baiyun Bridge is a steep single-arched stone bridge. To the north of the bridge is Yuyin highway (*lit.* the highway between Yuyao and Yin County); to the west is the towering Niushan mountain. It is about 50 meters away from Xiansheng Temple. Baiyun Bridge is 25.3 meters long and 3.8 meters wide, with a foundation of 1.1 meters high. The bridge arch spans 12.65 meters, with the rise of arch measuring 6.6 meters. The bridge has 22 stone steps on the north, and 24 on the south. The whole bridge looks like a rainbow over the river.

The design, the ornament and the sculpture of Baiyun Bridge are artistic. The stone bridge is tall and narrow, with high mountains on both sides and deep rapid currents under the bridge. It is just like an air corridor, or a rainbow over the river, impressively majestic. There are sixteen balusters on both sides of the deck, four of which are exquisitely and vividly carved with female and male lion heads at the top. There are also couplets on the side walls on both sides of the bridge arch. The one on the west reads, "Located on the boundary between Yin County and Yuyao County, the bridge sees people celebrating the prosperity there; linking families of Gong and Zheng, the village enjoys thousands of years of peace." The one on the east reads, "The rainbow of bridge crosses the river that links the north and the south; the moon lingers over the village that decorates the boundary between Yin County and Yuyao County." The characters in big size " 白云桥 " (Baiyun Bridge) are horizontally inscribed on the outer side of the bridge crown arch and the small characters " 光绪庚寅 " (the year of gengyin in the Guangxu reign) (1890) are inscribed on the right corner. The arch ring is composed of the stone slabs in parallel with longitudinal joints and stone arches are placed between balusters.

There is another bridge connected with Baiyun Bridge, called *tabu* bridge (*lit.* step bridge), which is located upstream at a higher position to the south of Baiyun Bridge. It is made of a line of rock steps laid in the stream at certain intervals that can be easily walked on. This is a common primitive stone bridge for mountain streams. It is also called "*yuantuo*" (a kind of turtle in the Chinese myth) because the rocks look like the back of turtles in the water when viewed from a distance. The bridge connects Baiyun Bridge and the southern bank of the river. Ancient skilled craftsmen combined the steep arched stone bridge and *tabu* bridge to considerably reduce the span of the arch ring, which not only saved labor and materials, but also facilitates flood discharge.

As the important landmarks of Yuyao City as well as the historical sites of city walls and gate towers in Yuyao, Tongji Bridge and Shunjiang Building stand next to each other, reflecting the layout of twin cities of Yuyao and the influence of Yaojiang River on the development of the city.

Tongji Bridge was originally built in the eighth year of the Qingli reign of the Northern Song dynasty (1048), initially named Dehui Bridge. After four times of ruins and reconstructions, the bridge was rebuilt into stone bridge in the third year of the Zhishun reign of the Yuan dynasty (1332) and renamed Tongji Bridge. The existing bridge, which was rebuilt in the ninth year of the Yongzheng reign of the Qing dynasty (1731), is a large, steep arched stone bridge with three arches and two piers, and is known as Bridge No. 1 in Eastern Zhejiang.

# Chapter Seven

# Yongfeng Warehouse: The Top Warehouse among Ancient Chinese Cities

Along the bustling West Zhongshan Road in Ningbo, within less than 2 kilometers from its intersection with Jiefang Road to Ximenkou, people can have a good view of the historical relics of different generations since the founding of Ningbo City in the Tang dynasty: Pagoda of Tianning Temple of the Tang dynasty, Yongfeng Warehouse of the Yuan dynasty, Moon Lake which flourished in the Song dynasty, Residence of the Fan Family of the Ming dynasty, the Drum Tower of the Qing dynasty, and so on. People might feel that they were walking through the time tunnel of the historical development of Ningbo City. The Site of Yongfeng Warehouse of the Yuan dynasty is located at the beginning of the tunnel.

On the Drum Tower, the Site of Yongfeng Warehouse is well within the sight. The ruins of buildings such as *yongdao* (path), column base, broken walls and drainage ditch of the Yuan dynasty are all telling people that a large area of buildings once stood here. Throughout the fog of history, one can imagine how busy and tense Ningbo was in its official warehouse in the Yuan dynasty as the starting place of the Maritime Silk Road and the city at the estuary of the Grand

Canal.

Ningbo City, or Mingzhou then, originally founded in the first year of the Changqing reign of the Tang dynasty (821), is a tiny Zicheng City located near the junction of Yuyao River, Fenghua River and Yongjiang River. It was built under the leadership of Han Cha, *Cishi* (prefectural governor) of that time. It was the location of the government office, and the political, economic and cultural center of Mingzhou in the Tang dynasty. Surrounded by water, Zicheng has a perimeter of 420 feet, extending from present-day Caijia Lane in the east to Hutong Street in the west, from Zhongshan Road in the south to the gate of Zhongshan Park in the north. Although small, Zicheng City is a relatively complete city with walls, gates and moats. The nowadays Drum Tower in Ningbo is the south city gate of Zicheng then. The walls of Zicheng were demolished during Yuan's occupation of Ningbo and were not reconstructed afterwards.

Luocheng City outside Zicheng City was built by Huang Sheng, *Cishi* of Mingzhou, in the first year of the Jingfu reign of the Tang dynasty (892). The pear-shaped Luocheng, with a perimeter of 2,527 feet, was laid out based on Mingzhou's natural water system. It was built along Yaojiang River in the north and Fenghua River in the east, and the Grand Canal in the west and south. According to archaeological calculation, Luocheng City might have 10 city gates. Compared with Zicheng City, Luocheng City was at least 20 times larger in area, and thus the greatly expanded city space laid the foundation for the ancient Ningbo City. After its establishment, it was destroyed and repaired several times over a long history until it was finally removed in the fall of 1931.

Zicheng City is the inner city while Luocheng City is the outer city. By investigating the early inner and outer cities of Ningbo, we can find not only the principles for ancient city construction, but also the distinct regional characteristics of Ningbo.

As the political area, Zicheng City is where the government offices were situated and state officials resided. Outside Zicheng City, Luocheng City, as the residential and commercial area, is where the common people lived. As a result, *yamen* (the headquarters or office of the head of an agency) where Zicheng is located was highly respected and protected. At the same time, the Drum Tower serves as the zero point of Luocheng City. Its upper or north part is the location of all administrative departments, and its lower or south part is the location of civilian and military forces. Its left or east part accommodates all the bustling streets, markets and public temples, and its right or west part with many academies had become the zone for book collection and education of Ningbo. The layout designs embody certain ideological characteristics such as "building inner city walls to defend the monarch and outer city walls to protect the people", "the temples to worship ancestors on the left while those to worship the god of land and crops on the right; the imperial court in the front while markets at the back."

The Site of Yongfeng Warehouse of the Yuan dynasty, located next to the Drum Tower in Zicheng City, Ningbo, was an important warehouse location from the Song dynasty to the Ming dynasty. According to historical records, in the first year of the Qingyuan reign of the Southern Song dynasty (1195), Mingzhou was changed into Qingyuan Prefecture, and a "Changping Barn" (the barn to stabilize the grain price) was built inside Zicheng City, "so as to keep a record of grain storage". In the Yuan dynasty, it was changed to "Yongfeng Warehouse". "Officials and clerks were sent to collect various items of confiscation, stolen goods, fines, and receipts, each reported to the higher authorities for examination." In the third year of the Hongwu reign of the Ming dynasty (1370), it was renamed Hongji Warehouse and "four cashier banks were designated respectively as Wen (literature), Xing (character), Zhong (loyalty), Xin (integrity)". Therefore, this site was important for the ancient warehouses of the Song, Yuan and Ming dynasties. In September 2001 and March 2002, Ningbo Institute of Cultural Relics and Archaeology carried out two rescue excavations of the site and finally found the foundations of two single buildings, including *yongdao* paved with bricks, courtyards, drainage facilities, wells and other cultural relics of the Yuan dynasty.

At the excavation site, the relationship of cultural layers is complex, and the superimposed ages range from the Han, Jin, Tang, Song, Yuan, Ming and Qing dynasties to modern times. During the excavation, the largest excavated relic is a large rectangular base of more than 1,300 square meters. The most completely preserved is the No. 1 single building, whose foundation is on that base. Upon the base of No. 1 site was also discovered a later building relic. Constructed with dark red stone blocks, it is preliminarily determined as Hongji Warehouse of the Ming dynasty.

Xu Pingfang, Chairperson of Archaeological Society of China and a well-known archaeologist, and Fu Xinian, an academician of the Chinese Academy of Engineering and a famous paleoarchitect, among others, believe that the Site of Yongfeng Warehouse, as the site of the largest single building of the Yuan dynasty found in China so far (the Yuan dynasty has a history of only 98 years, thus it has few historical remains), has its unique structure and is the only newly-discovered example of the ancient building structure in China. Among the historical and cultural sites protected at different levels in China, few can reflect the culture of the Yuan dynasty in Southern China. This site is not only the only architectural complex relic of the Yuan dynasty in Ningbo, but also the first discovery of such a large-scale warehouse site in China. It fills the gap in the archaeology of the Yuan dynasty and is of great significance to the study of the history of Chinese ancient architecture.

The wall foundation of Yongfeng Warehouse Site is 56 meters long and 16.7 meters wide, covering an area of 940 square meters. The surrounding walls are specially constructed in the way that square stone blocks with holes in the middle are closely arranged at the bottom of the wall, which form a rectangular building foundation. This type of ancient building structure had never been found before in China. Besides, such a large-scale single building is unique in the archaeological discoveries after the Tang dynasty in China.

There are three north-south partition walls in the middle of the foundation of the site, dividing the building into four large rooms, in which several square-hole stones are arranged regularly. Each of the room is divided into three smaller rooms according to the beam frame. All the walls center around a square-hole stone. It is presumed that on each square-hole stone a structural wooden pillar might have been erected, arranged closely with each other. On both sides are brick-wrapped walls in which broken brick pieces and earth are filled. The wall is 1.2–1.4 meters thick and no load-bearing pillars are found, so the wall has the load-bearing function. This type of building structure was documented, but physical examples had never been found before. Therefore, this is a newly found example of architectural structure at home and abroad, which is of great value to the study of Chinese architectural history.

In addition to two house foundations, the following were also discovered at the site: a rectangular brick platform that is 62 meters long from east to west and 21 meters wide from north to south, a brick path over 29 meters long and 6 meters wide in the northwest, an open brick ditch 118 meters long in total, a brick courtyard covering an area of 830 square meters, wells and moats, etc., all composing an interrelated and relatively complete structural area for government departments in the Song and Yuan dynasties. This is the most completely preserved ancient site found in the core area of Ningbo as a famous historical and cultural area.

In the warehouse, there are still many unsolved mysteries. Take the function of square stones with a square hole in the center for example. There is a presumption that it was used for wood or stone columns, but there is no wood chips in it, nor were any abandoned wood or stone columns excavated from the site. Another mystery is the even numbers of bays. In spite of the records and examples of buildings with even numbers of bays in early Chinese architecture, after the Sui and Tang dynasties, buildings, especially the official ones, generally have three, five, seven, nine bays or even more bays in odd numbers, which isi a practice all over the country. Why this single building base had four bays remains unknown. Besides, how can the ground part be restored? Where are the doors and windows? Researchers have made some assumptions for its restoration. Yet no final conclusion has been reached in the academic circle.

A large number of unearthed cultural relics, most of which are porcelain pieces, as well as a small number of coins of the Tang and Song dynasties, were also found while cleaning the remains. The unearthed porcelain products are a cluster of those made by five of the famous Six Kilns in the Song and Yuan dynasties. Yingqing porcelain (*lit.* shadow celadon porcelain) and white porcelain pieces made in Fujian account for more than half of them. In addition, pieces of the valuable Tang dynasty porcelain glazed with Persian peacock blue unearthed there made Ningbo the third city in China to have discovered such chips after Fuzhou and Yangzhou. It may also have been the hub for ceramics export on the "Road of Ceramics on the Sea", re-writing the oriental myth of "No market without Ningbo".

At the warehouse site, a private seal of the Tang dynasty engraved with " 文房之印 " (the seal of the literary chamber), and remnants of two steles of the Yuan dynasty were also found. The inscription on the steles includes the characters " 苫思丁 " (Shan Siding) and " 元帅府 " (Marshal's Mansion). According to historical documents, at that time, the Yuan dynasty classified the Chinese people into four groups: Mongolian, Semu (colored-eye) people, Han people and Southerners. Shan Siding was a Semu person. It is in line with the historical facts that, a semu marshal ruled Ningbo, which belonged to the "Southerners" area at that time. Besides, the unearthing of the private seal of the Tang dynasty confirms the popularization of school education and the emergence of literary celebrities in large numbers in Ningbo history.

# Chapter Eight

# Cicheng Complex of the Ming Dynasty: Zhejiang's Top Ming Architecture

Located 30 kilometers northwest of Ningbo, Cicheng is surrounded by mountains in three directions and a plain in the south. It is said that in the Spring and Autumn period 2,500 years ago, in order to commend to his later generations his achievements of destroying the State of Wu and being proclaimed Hegemonial Lord, Goujian, king of the State of Yue, built a city at present-day Wangjiaba in the southwest of Cicheng, which was known as "Gouyu" or "Gouzhang" in history. In the twenty-sixth year of the Kaiyuan reign of the Tang dynasty (738), *Caifangshi* (an official monitoring imprisonment and provincial and county government officials) Qi Huan petitioned the imperial court to set up Mingzhou City (Ningbo) at the east of Yuezhou, with the original location of Gouzhang as a county under Mingzhou. The imperial court appointed Fang Guan, grandson of Prime Minister Fang Xuanling, as the first county magistrate. Fang Guan visited the mountains and rivers of Gouzhang and finally chose the place where nowadays Cicheng was located as the site of county government. This county was built more than 80 years earlier than the capital of Mingzhou at the Three-river Junction. When

Fang Guan ascended Fubi Mountain in the north of the city and looked to the northeast at the towering Dongxiaozi Temple (the temple to honor the dutiful son Dong An) at the foot of Kanfeng Mountain, he was moved by Dong An's filial deeds of "collecting water for his mother", so he changed the county's name "Gouzhang" into "Cixi" (*lit.* stream of affection). Cicheng had been governed by Cixi County for more than 1200 years. It became Cicheng Town due to the modification of administrative division in 1954.

In Cicheng, an old city of over one thousand year's history, the ancient county *yamen*, examination hall and Confucius Temple that remain today highlight its time-honored history as a county. The ancient memorial arches, ancient bridges and aristocratic houses scattered everywhere in the city are telling the tradition of education in Cicheng. Since the Tang and Song dynasties, Cicheng had been widely known for its success in *keju* (the ancient imperial examination system), as in history it had 534 *jinshi* and five *zhuangyuan* (Number One Scholar, a title conferred on the one who came first in the highest imperial examination), nearly ten *bangyan* (a scholar who ranked second in the highest imperial examination), *tanhua* (a scholar who ranked third in the highest imperial examination), *huiyuan* (a scholar who ranked first in the metropolitan examination) and more than 1,200 *juren*. Especially in the Ming dynasty, there were a large number of talents in Cicheng. Cicheng (present-day Cixi), Yin County and Yuyao together were known as the "three golden counties" in imperial examination system with 902 *jinshi* altogether, which accounts for a quarter of the 3,458 *jinshi* in the whole province. In the imperial examination in the ninth year of the Zhengde reign of the Ming dynasty (1514), nine people passed the imperial examination in the same year in Cicheng, accounting for one sixth of all the *jinshi* of that year in Zhejiang. The county built an arch with the inscription "nine phoenixes flying together" to commemorate it. The celebrities or officials, living in Cicheng for generations or returning home after resignation, have built official residences and houses with local features, which have been passed down to this day, creating the most distinctive architectural culture of Cicheng today.

Taking a walk in Cicheng through the streets and lanes, you can see the historical culture of more than one thousand years everywhere. There are nearly 100 historical sites in the town that have been listed as historical and cultural sites protected at different levels or candidates for the titles. Among them, the folk houses of the Ming dynasty are the essence. According to the preliminary survey, in the chronological order of construction, there are Danai Hall and Fuma (emperor's son-in-law) Mo's Former Residence of the early Ming dynasty, Yao Mo's Former Residence, Jiadi Shijia Residence, *Fuzi Mentou* (*lit.* the gate with Chinese character " 福 ", which means blessings), Feng Yue's painted *taimen* (the building complex resided in by a clan with decent social status) and Buzheng's Residence (Feng Shuji was a *buzheng*, an administrative official, therefore his residence is also called Buzheng's Residence) of the Jiajing reign (1522–1566), and the Osmanthus Hall of the late Ming dynasty and early Qing period, so on and so forth.

Architecture is a solidified history, the tangible remains of history, and the concentrated embodiment of the national values, aesthetic interest and scientific development. However, due to wars and the changes of dynasties, it is also the buildings that are the most difficult to preserve, especially the buildings of timber structure. It is hard to find a well-preserved architectural complex of the Ming dynasty in other parts of our country. Cicheng, an area of less than four square kilometers, has retained so many folk houses of the Ming dynasty, which is quite rare in China. Therefore, it has become the historical architectural samples of the Ming dynasty researched by a large number of experts and scholars.

# I. Historical Representatives of Architectural Structures in the Ming Dynasty

## 1. Typical hall Buildings of the Early Ming Dynasty—Danai Hall and Fuma Mo's Former Residence

Danai Hall is located at No. 5, Sanmin Road, Cicheng, and its gate has long been destroyed. According to records, the gate of the Danai Hall faced south. There were a pair of *baogushi* on both sides of the gate, a stone stool on the left, and four round wooden gate stops above the gate. There is a plaque above the gate stops with the inscription " 甬东名阀 " (reputable family in the east of Ningbo) on it, and there is another plaque inside the door with the words " 诰封三代 " (conferment of honorary titles for three generations by imperial mandate).

Danai Hall is the Former Residence of Xiang's family in Cicheng. According to the genealogy of Xiang's family, the name began in the Song dynasty. As quoted by the genealogy from Chapter "Biography of Xiang Minzhong" in *The History of the Song Dynasty*: "At the beginning of the Tianxi reign（1017–1021）, Minister of the Ministry of Personnel was added to his official position... The emperor said that Xiang Minzhong was not disturbed at all by offers of official positions." It means that he was undisturbed by either honor or disgrace, able to bear solitude and to withstand promotion or reward. In the Song dynasty, Emperor Huizong wrote the words " 大耐堂 " (*lit.* hall of great tolerance) on a giant plaque, hence the name "Danai Hall".

According to *The History of the Song Dynasty*, Empress Xiang of Zhao Xu, Emperor Shenzong of Song (in reign from 1068 to 1085), was a great granddaughter of Xiang Minzhong. According to the folklore, Empress Xiang had no son, so she brought a concubine of Xiang's family into the palace as her own maidservant, who had been pregnant then and later gave birth to a son, Zhao Ji, who was claimed to be borne by Queen Xiang. After the death of Shenzong, Empress Dowager Gao supported Emperor Zhezong to succeed. Unfortunately, Zhezong died young without any offspring. Empress Dowager Xiang strongly proposed that Zhao Ji succeed to the throne and as a result, he became Emperor Huizong of the Song dynasty. Emperor Huizong of Song, whose original surname was Xiang, was a great grandson of Xiang Minzhong. When Emperor Huizong ascended the throne, he had "Danai Hall" built and wrote on a huge plaque to show his gratefulness to the Xiang family. Huizong honored the Empress Dowager Xiang as "Grand Empress Dowager Xiansu" and conferred honorary titles on three generations of his forefathers: Xiang Minzhong as King of Yan, Xiang Chuanliang King of Zhou and Xiang Jing King of Wu.

The existing Danai Hall has three bays facing south. It is a column-and-tie-beam timber structure, 13.87 meters wide and 11 *jie* deep, with five purlins in total. The building is tall and majestic. There are bracket sets on column tops and some components of the beam frame. The intercolumnar bracket set of the central bay comprises four bracket sets and is of six-rises-on-one-block style. The carved *tuofeng*, the *queti* in openwork carving, the beam ends with cloud patterns, the drum-shaped and disc-shaped column bases, and the walls made of reeds and coated with a mixture of grain husks and mud, all characterize the typical hall design of the early Ming dynasty. The east wing room with the hip roof are still preserved. The hall originally had more than ten plaques including one with inscription "Danai Hall" of the Tianshun reign of the Ming dynasty (1457–1464), which were all destroyed later.

In the history of Cicheng, there are three *fuma*, and Fufa Mo was one of them.

Fuma Mo is said to have moved from Yuyao to Cicheng. Both East and West Mojia Lanes are his former sites. According to *The Annals of Yuyao County*, Mo Shuguang served as *Zhongshu Sheren* (imperial secretary, mainly in charge of drafting imperial mandates at the Palace Secretariat) during the Shaoxi reign of the Song dynasty (1190–1194). His nephew Mo Zichun was *zhuangyuan* (the scholar who ranked first in the highest imperial examination) in the Bingchen year during the Qingyuan reign of the Song dynasty (1196). He was accepted as the son-in-law by Emperor Ningzong, Zhao Kuo, so he was called Zhuangyuan Fuma (Top Scholar Son-in-law). Mojia Lane is where they lived for generations.

The existing Fuma Mo's Residence is located at No. 25, Mojia Lane, Cicheng. It is a building of the Ming dynasty. The main house is still fairly complete. The main hall has five bays and two lanes, with eleven purlins. It is 20.6 meters wide and 12.4 meters deep. The eave columns have disc-shaped bases with slightly-cut angles. At the upper end of the middle column is a cross-shaped bracket set. The other columns use *pingpan dou* (flat block). Between the columns, some are one-step cross beams, while others are a combination of the one-step cross beam, dwarf columns and two-step cross beam. The front column is a small octagonal one, and the T-shaped *gong* is similar to the cicada-belly-shaped *queti*. The central bay has a mud floor, while the secondary and the side rooms have a wooden floor and a thin ceiling. About the second and third *jie* in the front of the central bay, there are five door panels under the two-step cross beams leading to the secondary rooms, which makes it different from Danai Hall. Behind the second column there are six door panels. And in the back *tianjing* courtyard there is a small garden.

The architecture of Fuma Mo's Residence is characterized by many columns standing erect on the ground, which is called *"zhuzhu shang"*(*lit.* column column up) and means "higher and higher, better and better", symbolizing auspicious implications. The lanes of the main building are connected with the corridors of the east and the west wing rooms. And there is a moon-shaped round door at the south end of the corridors leading to the front rooms. Seen from the general plan, the house is similar to the *"zoumalou building"*, in which people can walk under the roof in the house regardless of wind and rain.

## 2. The most Exquisite Wooden Building of the Ming Dynasty—Feng Yue's Dainted *Taimen*

There is a famous traditional opera in Ningbo, "Sanniang Teaches Her Son", which tells the story of Feng Yue, the owner of the painted *taimen*, and his stepmother.

Feng Yue's growth into a success in academic achievement and later into an official is considered to be attributed to his adoptive mother Sanniang (*lit.* the third daughter). It is said that when Feng Yue was young, his mother was told that her husband died away from hometown, and then died early in overwhelming grief. The young Feng Yue was then raised by Sanniang, his father's concubine. As a Ningbo proverb put it: "Stepmother's fist is as cruel as the sun in the sixth lunar month." Fortunately, Feng Yue's stepmother was more loving than her mother, encouraging him to study hard until he became a *jinshi*. In honor of Sanniang, there remains a well built in the Ming dynasty, Sanniang Well, next to Feng Yue's Former Residence.

In ancient time, *taimen* served not only as the shelter of the house, but also as a symbol of identity. The construction of the painted *taimen* was subject to strict examination and approval. Besides, it could not be built without the emperor's imperial permission and his direct appointment of carpenters and masons in the Ministry of Works to build it. Therefore, it was a great honor to return to hometown with such a privilege.

Feng Yue, with the courtesy name Wangzhi, a native of Cixi, is a *jinshi* in the fifth year of the Jiajing reign (1526). He has served as secretary of a Bureau of Nanjing Ministry of Works, vice director of the Ministry of Justice, Governor of Shuntian Prefecture (nowadays Beijing), Vice Censor-in-chief of the Court of Censors, and Minister of Nanjing Ministry of Justice. He resigned and returned home at the age of over 60. Having gone through the reigns of Jiajing, Longqing and Wanli, he was granted the plaque inscribed with " 三朝 伟望 " (Great Reputation Over Three Reigns). When he returned home after his completion of duty with moral integrity, Emperor Wanli granted him a painted *taimen* and had Wanjie Arch built for him.

Located at No. 2, Wanjiefang Lane, Taihu Road, the painted *taimen* faces south, with five bays, three central bays and two side bays, with a width of 13.16 meters and a depth of 7.05 meters in total. The ridge purlin is 5.8 meters high, with a front overhanging eave of 1.4 meters in width and a rear overhanging eave of 1.12 meters, both with the use of flying rafters. It has a gable roof and middle columns. One-step and double-step cross beams and two-step cross beams are also used. On the top of the double-step beam there is a cross-shaped bracket set to support the one-step cross beam. Square blocks are applied to each column top to make a cross-shaped bracket set. The horizontal *gong* is mostly *chonggong* (double-tier bracket-arm supporting a joist). Under the beams are T-shaped *gong*; the three in the front are trusses and the four at the back are purlins. There are wing walls in front of the side rooms. The upper part of the wing wall has *queti*, which is built with obliquely arranged, smooth-facing square bricks, and the lower part is a stone Sumeru base. There are also wood carvings of dragon, phoenix, kirin, ganoderma lucidum and *ruyi* in openwork carving on some bracket sets and columns. All the beams, columns, tie-beams, architraves and bracket sets are decorated with color-painted patterns such as "peacock and peony", "crane" and "lotus leaf". The paintings are so exquisite that they are rare and precious in China.

### 3. The Most Glorious Building of the Ming Dynasty—Jiadi Shijia Residence

The name "Jiadi Shijia" dates back to the Jiajing reign of the Ming dynasty (1522–1566). According to the records of *The Annals of Cixi County*, Qian Zhao, the owner of the house, passed the provincial-level imperial examination in the seventh year of the Jiajing reign (1528), and then the highest imperial examination as a *jinshi* in the eleventh year; he worked as *qianshi* (an official in charge of official business). His son Qian Weiyuan, a *zhusheng* (alias *xiucai*, referring to those who passed the county-level imperial examination) took care of his grandfather until his decease. His grandson Qian Wen was recommended as a *jinshi* in the thirty-fifth year of the Wanli reign of the Ming dynasty (1607).

His several other descendants passed the imperial examinations too. Therefore, the Residence of Qian's family became known as Jiadi Shijia Residence. The plaque bearing the four characters inscribed by Wen Zhengming had once been placed inside the original *taimen*, which was later destroyed.

Just as the family of Qian Zhao is the role model of scholars due to its successful candidates in imperial examinations among three generations, the unique architectural style of its residence constitutes a special architectural landscape of the ancient residence.

Facing south, the whole residence covers an area of 1,863 square meters with a construction area of 1,360 square meters. The plan layout is a longitudinal rectangle, adjacent to *Fuzi Mentou* in the east. The main building on the central axis has two *jin*, one in the front and the other at the back. The house is composed of *taimen*, the second gate, the front hall, the rear hall, and the left and the right wing rooms. *Taimen* is located in the southeast corner. It is a brick and wood structure. The front hall has five bays. The central bay has six beams. The beams and columns are thick with oval cross-sections. The *qunban* of *jigua* columns and dwarf columns, which are two types of the king post, are in the shape of a round tongue. The columns have entases. Two dwarf columns are used to support the front and rear upper intermediate purlins on the two-step cross beams in the secondary rooms. There are bracket sets at all the column tops and the rear middle hypostyle columns. The front hall and the rear hall are of single eave and gable roof. A wing building with a hip roof is built at both of the south ends of the left and the right wing rooms.

What is the most special about it is that *hudou* (the supporting bracket) is made into the square shape with rounded corners, also the so-called "*xuedou*" (*lit.* boot bucket) in *Building Standards* of the Song dynasty. *Hudou* has an obvious inheritance relationship with *douqi* in the Great Hall of Baoguo Temple of the Song dynasty in Ningbo and in Yanfu Temple of the Yuan dynasty in Wuyi County, Zhejiang Province. The house, from the plan layout to the architectural form, has characteristics of Ming dynasty residential buildings and reminiscence of Tang and Song architecture, and therefore is a typical building that many experts and scholars must refer to when studying residential buildings in the Ming dynasty.

## 4. Other Buildings of the Ming Dynasty

Among the representative historical buildings of the Ming dynasty are *Fuzi Mentou*, Yao Mo's Former Residence, Ancestral Hall of Liu's Family, Osmanthus Hall, Buzheng's Residence and Feng's Residence.

*Fuzi Mentou* was originally the residence of Feng Shuji, Provincial Administration Commissioner of Huguang Province in the Wanli reign of the Ming dynasty (1573–1619). Through the dynasties of Ming, Qing and the period of the Republic of China, it looks a little dilapidated and the walls have been invaded by weeds, but it still vaguely retains some traces of its past glory, as if telling its stories of the past. Feng Shuji was a *jinshi* in the thirtieth year of the Jiajing reign of the Ming dynasty (1551). He purchased the house and built walls surrounding it. A screen wall was added in front of the original secondary gate, and another gate was added outside. The character " 福 " (*lit.* blessings) was also engraved in openwork carving on the screen wall (now destroyed), which is the origin of the name "Fuzi Mentou".

Looking around the whole courtyard, the secondary gate no longer exists; what remained are only the screen wall, the main gate, the front hall, the middle hall, the left and the right wing rooms and the back building. With a rectangular layout, the house covers an area of 1,089 square meters, with construction area of 1,086 square meters.

Walking around, you may feel the touch of ancient humanistic atmosphere. Perhaps Feng Shuji was inspired by Beijing *siheyuan* (courtyard house with a fully enclosed courtyard), for the whole layout is very similar to that of northern houses. The symbolic screen wall of "Fuzi Mentou" is a masonry structure facing north. Its carved stone base remains intact, which is in the style of Sumeru base and carved with clearly visible grass patterns. The ancient screen wall has multiple functions, such as for embellishment, revealing identity, symbolizing power and wealth, privacy and so on. It balances the whole architectural layout, for placing a shelter at the entrance is in line with the practice of ancient Chinese architecture and reflects the implicit and introverted characteristics of ancient Chinese people. Moreover, the character "*fu*" on the screen wall embodies the owner's hope that the house could bring good luck to his family.

In addition to the intact screen wall, the front hall with a gable roof of single eave covered with black tiles, has the architectural style of the Ming dynasty. It has five bays, including three central bays and two side bays. Each of the front eave columns of the central bays have eight slightly-cut angles and a square column base. The other column bases are flat bead shaped, and the column tops are designed with obvious entasis. The column-beam-and-strut structure combines the one-step cross beam, dwarf column and two-step cross beam. The gap between beams is filled with reeds, which were wrapped in a mixture of grain husks and mud.

Yao Mo's Former Residence has lost its original scale. What still remains are three bays and two lanes in the back courtyard, which have double eaves and nine purlins made of thick timbers. The top of the middle columns is equipped with a cross-shaped bracket set. The eave column has slightly-cut angles and disc-shaped column base. The other column bases are drum-shaped. The walls have woven bamboo as the core and are plastered with the mixture of grain husks and mud.

Yao Mo (1465–1538), with Yingzhi as his courtesy name, born in Cixi, was a *jinshi* in the sixth year of the Hongzhi reign. He first served as Secretary of a Bureau of Ministry of Rites, then was promoted to vice director of the ministry, and later further to *tixue qianshi* (an official in charge of education) in Guangxi. His son, Yao Lai, with Weidong as his courtesy name, was the first place in the imperial examination in the second year of the Jiajing reign of the Ming dynasty (1523), and was offered by the emperor to work in Hanlin Academy. He was once summoned to compile the book *Minglun Dadian (A Great Dictionary of Rites in the Ming Dynasty)*, but he rejected it. Later, he was promoted as a *shidu xueshi* (a scholar in charge of the proofreading of *zouzhang*, the memorial to the emperor).

Ancestral Hall of Liu's Family and Osmanthus Hall are located in the area where the Liu's family lived in the Ming dynasty. The Liu's family was originally a big family in Cixi, whose ancestors were Liu Mian, *tai-chang-si-cheng* (assistant of minister of ceremonies) in the Southern Song dynasty, whose father is Liu Chun'an. Four generations of the large family lived together. According to the records of *The Annals of Cixi County*, Liu Mian's Residence in the Song dynasty was named Shicai Hall (*lit.* generations cheer hall). Given that the tradition of scholarship and official titles have been kept through the generations of Liu Mian's ancestors for hundreds of years, the city governor named his hall "Shicai Hall" on his 90th birthday. Yet the plaque of "Shicai Hall" of the Song dynasty was later destroyed.

Ancestral Hall of Liu's Family is built in the Ming dynasty for Liu Mian's descendants. The existing hall has three bays, with a rectangular plan and a gable roof. The beam frame is tall and it is nine *jie* deep, using ten purlins. The central bay adopts the column-beam-and-strut structure, and under the five purlins are *dingtou gong* and intercolumnar bracket set. The middle bay has three bracket sets, while the secondary rooms use two sets. The lower edge of the dwarf column is in the shape of a round tongue. The columns have with a drum-shaped base. At the northeast corner of the gable wall stands a stele with the inscription of "In accordance with the government decree, the Liu's family owns this property forever and the whole family are all exempt from the corvee"; the stele was erected in the ninth year of the Jiajing reign of the Ming dynasty (1530). The ancestral hall is one of the earlier constructed in the Ming dynasty in Ningbo.

Osmanthus Hall was rebuilt in the forty-eighth year of the Wanli reign of the Ming dynasty (1620). It was named Osmanthus Hall because of the previous "Liangui Fang" (*lit.* arch of connected osmanthus) at the gate, osmanthus trees planted next to the house, and most importantly, the owner's love for osmanthus and poetry.

This residence is a *siheyuan* with a gable roof, the gate facing south. It is composed of *daowu* (the south room), front hall, sitting room, back building, left and right wing rooms, and a well and a pool in the back. On the east side of the *daowu* and the south of the east wing room is the *taimen*. There used to be a pair of stone lions in front and a screen wall directly at the very north of the *taimen*. Now *taimen* and the screen wall at the front hall have been destroyed. The back building and the west wing room have been alternated to varying degrees. Only the sitting room, three bays wide, has preserved its original appearance. The front eave column, small and octagonal, is equipped with a cross-shaped bracket set and a disc-shaped column base. T-shaped *gong* is used under the beams. The partition wall is made of reeds and wood strips and wrapped with a mixture of grain husks and mud. There are still several eaves-end tiles with parallel sawtooth patterns remaining on the back eaves, which is characteristic of the architectural style of the Ming dynasty.

# II. Features of Cicheng Architecture of the Ming Dynasty

The earliest extant remains of residential buildings in China are those of the Ming dynasty. The Ming dynasty was ruled by the Han landlord class, established on the basis of the peasant uprising at the end of the Yuan dynasty. In the early Ming dynasty, the Ming regime, in order to consolidate its rule, implemented a variety of measures to develop production and made progress in construction technology. Through investigating the Ming dynasty architecture in Cicheng, we can sum up their characteristics as follows.

Bricks have been widely used in building walls of folk houses. Before the Yuan dynasty, despite the existence of brick pagodas, brick tombs, brick arches of waterways, etc., the timber frame buildings mainly used earthen walls, while bricks were only used in paving the ground and building the platform foundation and wall foundation. Brick walls had not been widely used until the Ming dynasty. Due to the common use of hollow walls in the Ming dynasty, the required number of bricks was reduced and therefore the brick wall was popularized. This paved ways for the development of buildings with gable roof. In the Ming dynasty, the processing technology and quality of bricks were improved. For instance, the wing walls in the shape of Chinese character " 八 " in front of the side rooms of Feng Yue's painted *taimen*, with a length of 2.4 meters on each side, use smooth-facing bricks that are obliquely arranged, with *queti* at the top. It can be seen that Ming dynasty witnessed the maturity of brick processing and carving technologies. Brick processing is known as "*zhuanxi*" (fine brick work) or "*zhuanzuo xizuo*" in China, which means to process brick surfaces and edges with a plane to achieve the extremely flat surfaces and straight lines.

Second, in terms of timber structure, after its simplification in the Yuan dynasty, a new framework was formed and finalized in the Ming dynasty, where the structural function of bracket set was reduced, the structural integrity of beams and columns was strengthened, and the entasis of components was simplified. Although these trends had emerged in some buildings of the Yuan dynasty, they have not been as popularized and stable as in the Ming dynasty. *Douqi* on the column tops no longer played an important structural role as in the architecture of the Song dynasty. The *ang*, originally used as an inclined beam, had also become a purely decorative component. Therefore, the official-style architecture of the Ming dynasty is distinguished from that of the Song dynasty in that it looks more rigorous and steadier, but it is not as open and lively as that of the Tang and Song dynasties. Due to the common development of local folk buildings and the corresponding improvement of technical skills, the technical book in carpentry titled *Luban Construction Treatise* was composed, which provides valuable data on folk houses and furniture in the Ming dynasty. For example, the bracket set of the Jiadi Shijia Residence had no structural function and was only used as a symbol of the specification and hierarchy of the residence, which had a great impact on the folk houses in the

Qing dynasty.

Third, in ancient times, residential houses were not only the place to live in, but also the symbol of the owner's identity. In the early Ming dynasty, the width and depth measured in the number of bays of official residences were obviously restricted. Officials were not allowed to use gable-and-hip roofs or double eaves roofs, *chongqi* or caisson ceilings. These restrictions had been originally set for common people in the Song dynasty, but in Ming dynasty they were for the officials as well, which meant that except for the royal family members, no matter how high one's official position was, he could only use overhanging gable roof and gable roof, rather than the gable-and-hip roofs. In addition, the residences of aristocrats and officials were divided into four hierarchical levels and strictly restricted in terms of the gate, the number of bays in the hall, the depth, and the paint color. According to the regulations of the early Ming dynasty, Danai Hall could have three large bays at most, even though it is tall and grand.

Fourth, the specifications also cover the architectural details. Take the column base for example. There are two types of column base then—drum-shaped and felt hat-shaped. The former has a maximum diameter at less than half its height; the latter looks thicker and shorter, with an edge at the lower part, a cylindrical upper-part, and a slightly bulging lower part. The column bases of the Ming dynasty had an arc-shaped transition from the edge to the upper part, while those of the Qing style had an angled transition and mostly used the red stone as the material. The partitioning of the beam frame mostly used reed stalks and thin bamboo chips as the core and was covered with a mixture of grain husks and mud, commonly known as mud walls. The *tianjing* courtyard is paved with bar-shaped stones, while the hall floor is mostly compacted tabia, which is solid and durable.

Wandering around the buildings of the Ming dynasty in Cicheng, people will find that despite the alternation and dismantlement of some *taimen*, wing rooms and back buildings, the halls, the major part of the residences all retain their original structure of the Ming dynasty. This provides some clues for people to study the evolution of these buildings. The hall is generally an important place for a family to receive guests, discuss family business, and worship ancestors and gods. It serves as a symbol of prosperity of a family. Therefore, with a prudent attitude towards the "old house" handed down by the ancestors, people did not rebuild it rashly, but worked to add more glory to it. For example, Danai Hall, built after the descendants of Xiang Minzhong (Steward-bulwark of State during the reign of Emperor Zhenzong of the Song dynasty) moved to Cicheng, still kept sixteen or seventeen plaques at the founding of new China, among which the earliest plaque with the inscription " 大耐堂 " (Danai Hall) was a relic from the Tianshun reign of the Ming dynasty (1457–1464). Besides, the hall was carefully designed in construction in strict accordance with the building regulations of the Ming dynasty. Also, the choice of materials and decorations were so exquisite that they were not vulnerable to artificial or natural damage. The folk houses of the Ming dynasty, such as Dafang Yuedi and Fan's Residence in the urban area of Ningbo, are similar cases. This is the main reason why these folk houses could survive even four or five hundred years.

# Chapter Nine

# Ancient Academies Through Thousands of Years

The lingering scent of books throughout Ningbo history attracts us to the world of the ancient academies.

Chinese academies originated in the Tang dynasty. Yuan Mei of the Qing dynasty wrote in his *Essays in Suiyuan*: "The name of the academy dated from Emperor Xuanzong of the Tang dynasty. Lizheng Academy and Jixian Academy were both set up in the imperial court." Their initial function was to compile and proofread books. The academy with educational function emerged in the late Tang dynasty and Five Dynasties period. It was an important place for private gathering and lecturing. According to the documents recorded in *The Annals of Ningbo*, the earliest academy in Ningbo history was Penglai Academy set up by Yang Hongzheng, a magistrate of Xiangshan County, in the Qixia Taoist Temple at the foot of Penglai Mountain in the northwest of the county in the fourth year of the Dazhong reign of the Tang dynasty (850). The academy was funded by the school-owned land and mainly taught Confucian classics.

The Song dynasty was a period of rapid development for Ningbo academies. Yang Shi, Du Chun, Wang Zhi, Wang Yue, and Lou Yu, the "Five Masters of the Qingli period" in the Northern Song dynasty, once set up academies to give lectures on classics and history. The famous ones are Zhengyi Lougong Lecture House started by Lou Yu and the Taoyuan Academy by Wang Yue, to which a plaque was awarded by Emperor Shenzong of the Song dynasty

The "Four Masters of the Chunxi period" in Mingzhou in the Southern Song dynasty, Shu Lin, Shen Huan, Yang Jian and Yuan Xie, inherited Lu Jiuyuan's theory and formed Siming School with Gao Kang and others. They gathered in Mingzhou and set up academies to give lectures. The more notable ones are Zhuzhou Three Scholars Academy, Master Yang Wenyuan Academy, and Chengnan Academy.

In the Ming and Qing dynasties, Yaojiang culture and Eastern Zhejiang culture in Ningbo's academic history flourished, with Wang Yangming and Huang Zongxi as representatives, who gave lectures in academies around the country. The renowned academies in the Ming dynasty include Zhongtian Pavilion, Yaojiang Academy, Jingchuan Academy, etc.; those of the Qing dynasty include Yongshang (Ningbo) Zhengren Academy, Moon Lake Academy, Yucai Academy, etc.

During the Guangxu reign in the late Qing dynasty, there emerged academies founded by foreign missionaries in Ningbo, such as Trinity Church School on Xiaowen Street, Frederick Academy at the Old Bund, Huaying Academy near Zhangbin Bridge in Jiangdong District.

Statistics show that since the Tang dynasty, there have been more than 100 well-known and well-documented academies in Ningbo, which became a lecturing centers in Zhejiang Province. As for today, most of the academy buildings have been destroyed. What have been preserved are only Zhongtian Pavilion where Wang Yangming once lectured, Yongshang (Ningbo) Zhengren Academy, or Baiyun Manor, Yuying Academy, Jinshan Academy, the relics of steles and drum-shaped stone relics in Qiushan Academy.

Among them, Zhongtian Pavilion and Yongshang (Ningbo) Zhengren

Academy are the best known.

On Longquan Mountain in Yuyao City, there is Zhongtian Pavilion, where Mr. Wang Yangming used to give lectures. Zhongtian Pavilion was initially built in the Five Dynasties period and became a part of Longquan Temple in the Ming dynasty. Wang Yangming had lectured here for several times, among which two times were definitely recorded in historical documents. One was in the sixteenth year of the Zhengde reign of the Ming dynasty (1521), when Wang Yangming returned to Yuyao to visit his ancestral tombs and gave lectures to more than 70 people, including his disciple Qian Dehong, in Zhongtian Pavilion. The other was in the fourth year of the Jiajing reign of the Ming dynasty (1525). "The lectures are scheduled on the dates of *shuowang* (the first and fifteenth day of each month), and the eighth and the twenty-third day of each lunar month in Zhongtian Pavilion of Longquan Temple", and the number of disciples present reached more than 300 at its best. Wang Yangming also set up rules for them in "Guidelines for Zhongtian Pavilion Disciples", and wrote them down in person on the wall to regulate and encourage them. At the age of 57, Wang Yangming even wrote a letter to "ask about the lectures of Longshan Mountain in Yuyao" when he was seriously ill after his successful attack on Duanteng Gorge, Bazhai in Guangxi. It can be safely concluded that Master Wang has an unusual relationship with Zhongtian Pavilion. Unfortunately, the pavilion was later destroyed in war. In the twenty-fourth year of the Qianlong reign of the Qing dynasty (1759), Liu Changcheng, then magistrate of Yuyao County, established Longshan Academy. He employed teachers to give lectures to and evaluate disciples every year. Wang Yangming's memorial tablet was laid upstairs, while children disciples took classes downstairs.

The existing Zhongtian Pavilion, as a Historical and Cultural Site Protected at Municipal Level of Yuyao, was rebuilt in the fifth year of the Guangxu reign of the Qing dynasty (1879) and repaired in 1985. Stepping into the main hall of Zhongtian Pavilion, we can see a portrait of Master Wang Yangming in the middle, with a wide forehead and high cheekbones, thin and serious-looking. It was him, a philosopher, who broke through Zhu Xi's Neo-

Confucianism (*lixue*), which was full of defects and rigidness, and surpassed the former sages with the philosophical proposition of *zhiliangzhi* (extension of innate knowledge of the good) and developed *xinxue* (School of Mind) to its highest level.

Yongshang (Ningbo) Zhengren Academy is located in Baiyun Manor in the west of Ningbo City. It is a simple and solemn ancient building with black bricks and walls. It was once the place where the Confucian master Huang Zongxi gave lectures in the late Ming and early Qing period.

Baiyun Manor was originally owned by Wan Tai, Secretary of a Bureau of Ministry of Revenue in the late Ming dynasty. Later, it got its name Baiyun Manor because his son Wan Sixuan was named "Mr. Baiyun" in the wake of his writing *The Collection of Baiyun*. Walking through the *yimen* gate, we come to the place where Huang Zongxi gave lectures. There is a portrait of him on the front wall of the front hall. Wearing a Confucian scarf (a sort of headwear), he sat sideways, staring into the distance, as if in contemplation. There remain eight seats in the hall, together with a tea table, an Eight Immortals table (a square table for eight people) and an altar, bamboos swaying outside the window. The rear hall has been used for the exhibition room of Huang Zongxi's life story, which demonstrates the sage's bumpy and extraordinary life.

Huang Zongxi was also known Mr. Lizhou (his literary name), with the courtesy name Taichong and the assumed name Nanlei. He was a representative of the Eastern Zhejiang School in the early Qing dynasty. In the sixth year of the Kangxi reign of the Qing dynasty (1667), he reopened Zhengren Academy in Shaoxing, which was originally founded by Liu Zongzhou, a Shanyin (i.e., Shaoxing) scholar when he was giving lectures there. Huang Zongxi studied from Liu Zongzhou in his youth. When the Ming dynasty collapsed, Liu Zongzhou went on a hunger strike and died. Huang Zongxi carried on his will. In the seventh year of the Kangxi reign of the Qing dynasty (1668), at the invitation of his friends in Ningbo, he went there to give lectures and organized Zhengren Lecture and made a difference in the local literary style. Originally, the lectures were given in Wan Tai's[1] Residence at Guangji Street. Later, it moved to Yanqing Temple, and finally to Baiyun Manor. Since then, Baiyun Manor had become the academic center of Eastern Zhejiang School. During the Qianlong reign (1736–1795), Quan Zuwang, a disciple of Huang Zongxi's private school and the "last monument" of Eastern Zhejiang School, added two characters " 甬上 " (Ningbo) to the name in order to distinguish it from the "Zhengren Academy" in Shaoxing, and hence the name "Yongshang Zhengren Academy". Yongshang Zhengren Academy and Baiyun Manor were damaged in the late Qing dynasty. In 1934, when Yang Yicheng and other Ningbo people visited the site of the academy, they raised funds to have it restored. Since the representatives of the Eastern Zhejiang School such as Huang Zongxi, Wan Sitong and Quan Zuwang had all lectured here, there have been an endless stream of scholars coming for a visit.

Yuying Academy is located in Longgong Village, Shenzhen Town, Ninghai County. At the end of the Northern Song dynasty, Chen Zhongliang moved from Pinghu, Xinchang to Longxi. His descendants started to build the ancestral hall in the Ming dynasty. In the early Qing dynasty, due to the proliferation of the clan members, another ancestral hall was built in the west of the village, called "Chongde Hall". Because the local people in the Longgong Village attached great importance to education, they built Wenchang Pavilion, a free private school built at the top of Yuelong Bridge, as early as in the early Qing dynasty. After several times of being destroyed and reconstructed, in the late period of Republic of China, since the old school buildings could not accommodate the large number of students, it was moved to Chongde Hall that was later slightly reconstructed. With the environment getting better, it was renamed Yuying Academy.

The overall layout of the academy faces south, with a wall around the front yard in the south. There is a gate in the east, and there used to be a main gate at the front hall, but only the west gate is left. Along the central axis

---

[1]    Wan Tai, officer of the Ministry of Revenue in the late Qing dynasty.

from south to north, there are Wufeng Building (*lit.* five-phoenix building), the *tianjing* courtyard, and the main hall. It covers a total area of 539 square meters.

The main hall faces south, with a single eave and a gable roof covered with tiles in the way of yinyang hewa[1]. The total width is 14.19 meters and the total depth is 10.11 meters. It is a mixture of the column-beam-and-strut structure and the column-and-tie-beam structure. The southern bays have seven purlins, two-step cross beams at the front and back, and five columns. The eave columns adopt column-top bracket set. There is a projection of the false *ang* on *zuodou* (the largest bearing-block in a bracket set), with a variant form of the bracket set being used. The huge *queti* is carved with cloud patterns.

It is a two-story structure covered with *yinyang* roof tiles, with three wing rooms in the east and west respectively. It has a total width of 11 meters and a total depth of 9.65 meters. The balustrades are turtle shaped and carved with patterns. On the upper floor of the front hall are five small screens with the pattern of " 卍 ".

The front hall is a two-story structure with roof tiles arranged in the way of *yinyang hewa*. It has a width of five bays, 22.81 meters in total, and a total depth of 5.66 meters, with a round ridge roof.

Very few academies have been preserved in Ninghai County, and even fewer is so completely kept as Yuying Academy Rebuilt from the ancestral hall, it retains the look and decorative art of the ancestral hall, as showcased in its balustrades with lattice patterns at the two wing rooms and the second floor of *daozuo*, which are characteristic of the local architecture in Ninghai and have certain technological value.

Located in Shipu, Xiangshan, Jinshan Academy was initially founded by four local brothers, one of whom Xu Chao, in the eighth year of Daoguang reign of the Qing dynasty (1828). It was then named "Chongde Yishu" (Chongde

---

[1]  A popular method to interlock the barrel tiles, by placing rows of cupped tiles on the roof, with rows of arched tiles spanning between them.

Free Private School). As recorded on the stele of Chongde Yishu, "Xu Chao, with Zhuoxuan as the courtesy name, together with his brothers Xu Jiongzhai, Xu Dengwu and Xu Yanguan, founded the free private school to the north of Shicheng in the Wuzi year of the Daoguang reign (1828), and it began to take shape." By the thirteenth year of the Tongzhi reign (1874), the academy had declined. Yang Diancai, Associate Administrator, negotiated with Xu's descendants and reported it to the higher-level officials. After revision, it was renamed as "Jinshan Academy" and then teachers were invited to give lectures day and night. After that, the nearly decaying Chongde Yishu was revived in the name of Jinshan Academy, which was actually the initial development of schools in Shipu.

In the thirty-second year of the Guangxu reign (1906), in the wave of "Abolishing the Imperial Examination and Practicing New Schooling", Jinshan Academy was renamed "Jingye Primary School for Senior Classes". In 1925, it was again renamed "District-owned Jingye Combined Primary School". So far, the name of the school developed from *yishu* (free private school), *shuyuan* (academy), *xuetang* (*lit.* learning hall) and finally to *xuexiao* (*lit.* learning organization).

After the founding of the People's Republic of China, it was renamed "Shipu Primary School". The famous historian Wu Han once studied here. The Jinshan Academy still has the main room of "Jingye Hall" and a five-bay back building. In 2002, it was restored based on the historical architectural style in the reigns of Daoguang and Tongzhi. Later, it was established in *The Protection Planning of Shipu Historical and Cultural Reserve* as "Shipu Education History Exhibition Hall".

Qiushan Academy is located in the Central Elementary School of Xianxiang Town, Yinzhou District. There are only stele relics and drum-shaped stones from the Xianfeng reign of the Qing dynasty (1851–1861). The two steles are of the same size, both 2.5 meters high, 1 meter wide and 0.12 meter thick. On top of the first one are four characters in large regular script "球山碑记" (Qiushan stele inscription). The text arranged in four columns describes the situation when Qiushan Academy was first built. The names of the donors and the amount of the donated land or money are engraved after the text. On top of the second one are two characters " 碑记 " (stele inscription), with its content following the first one. What is engraved on it are also the donors' names and the amount of land or money they donated. The inscription of the lower part reads "Erected by Lin Zhongyue in the sixth lunar month in the eighth year of the Xianfeng reign, the stele erected by directors Zhu Zhaoshen and Zhu Xingzheng". A pair of drum-shaped stones of the same size and style are well-preserved in the Academy; the drums with the surfaces on two sides are 1.3 meters high, 1.55 meters wide, and 0.80 meters in diameter.

Qiushan Academy was founded by Zhu Zhaojia, Zhu Xingzheng, among others, by raising more than 300 acres of school land and more than 4,000 strings of coins. It was established in the fifth year of the Daoguang reign (1825) and completed in the sixth year (1826). It was originally sited to the west of Yangjia Bridge in Xianxiang, covering an area of three acres. The three sections of school buildings face south. The Kuixing Pavilion in the front section is a timber structure with upturned eaves and protruding corners. The auditorium in the middle section has five bays, which is a mixture of the column-beam-and-strut structure and the column-and-tie-beam structure. The building in the back section has five bays, two lanes, and five accessory flat houses. Its scale and facilities were second to none in the county at that time.

# Chapter Ten

# Private Libraries Through History

Beside the well-known Tianyi Pavilion, few other existing private libraries in Ningbo are known to people. However, these libraries are an important part of book collection culture in Ningbo.

As recorded by Chapter "Documentary Annals" in *The General Annals of Yin County*, since the migration of the Song dynasty to the south of China, with so many aristocratic families in Ningbo, library buildings, the rise of the book printing industry, and the prevalence of book collection, there emerged quite a few private libraries in Ningbo. According to *The Annals of Ningbo*, after the Song dynasty, there were nearly 80 famous book collectors and more than 40 well-known and verifiable private libraries in Ningbo, which makes it one of the centers of book collection in Zhejiang Province.

So far, many private libraries have disappeared, with only a few standing the test of time. They are Tianyi Pavilion, Fufu Room, Yanyu Building, Shuibei Pavilion and Woji Hut. These scattered private libraries are like pearls, each shining with the brilliance of the city's book collection culture. These ancient library buildings bear witness to the rich cultural memory of the city and constantly nourish it with a distant sense of poetry.

# I. Fan Qin's Tianyi Pavilion

Fan Qin, with the courtesy names Yaoqing and Anqing and the assumed name Dongming, was born on the nineteenth day of the ninth lunar month in the first year of the Zhengde reign of the Ming dynasty (1506) and died on the twenty-eighth day of the ninth lunar month in the thirteenth year of the Wanli reign (1585). He became a *jinshi* in the eleventh year of the Jiajing reign (1532). Initially, he was appointed Sub-prefectural Magistrate of Suizhou, Huguang Province. Later, he was promoted to Vice Director of the Ministry of Works responsible for such projects as construction and maintenance. In the thirty-seventh year of the Jiajing reign (1558), he served as Provincial Administration Commissioner of Henan Province and was later promoted to Vice Censor-in-chief. Two years later, he was promoted to the Minister of War. In the tenth lunar month of the same year, he resigned from his official post and returned to his hometown. He loved books all his life so much that he collected books everywhere he went, especially works of contemporary people. As a result, among his collection of books, many are local annals, political books, memoirs and poetry collections of the Ming dynasty. His library was originally named Dongming Cottage after his art name. Later, with an increasing number of books, the old house became not big enough, so a new building, Tianyi Pavilion was built to house them.

Tianyi Pavilion is also called "Baoshu Pavilion" (*lit.* treasure book building). This is a building with a gable roof of double eaves and double stories, with a total height of 8.5 meters. The ground floor is six-bay wide and six-bay deep, with corridors in front and back. The second floor, except for the staircase, is a large room facing south, separated by bookcases, with windows in front and back for ventilation. There are six parallel rooms downstairs and one room upstairs, which means "Heaven is one" and "Earth is six". In addition, Tianyi Pool was dug in front of the building and connected to the

Moon Lake, serving not only for beautification but also for fire prevention.

It is said that one day, while reading the words in *The Book of Changes* that "Heaven is one, and it gives birth to water; earth is six, and it gives shape to it", Fan Qin was inspired to design this building based on it, and renamed Dongming Cottage as "Tianyi Pavilion". The complex of Tianyi Pavilion Buildings is in good order and simple style. The patterns of caisson ceiling are water patterns and ancient water animals, symbolizing the prevention of fire by means of water. As a model of private library, Tianyi Pavilion has a far-reaching impact. The seven well-known libraries which Emperor Qianlong ordered to keep the book *Complete Library of the Four Treasuries*, including Wenyuan Pavilion in the Imperial Palace, Wenyuan Pavilion in Yuanmingyuan Park, Wenjin Pavilion in Chengde, Wensu Pavilion in Shenyang, Wenhui Pavilion in Yangzhou, Wenzong Pavilion in Zhenjiang, and Wenlan Pavilion in Hangzhou, were all designed and constructed in imitation of Tianyi Pavilion in style and structure, bringing it the big fame.

In the early Qing dynasty, Fan Guangwen, a great grandson of Fan Qin, invited a reputed craftsman to build rockeries near the Tianyi Pond, surrounding it with bamboo and wood. The artificial hills are exquisitely designed into the shape of "nine lions and one elephant", "the shepherding old man", "a beauty looking into the mirror" and three Chinese characters " 福、禄、寿 " (happiness, wealth and longevity). The rockery animals and figures are true to life. A rock that looks like a young girl gazing at the library building is said to be based on Qian Xiuyun, niece of Qiu Tieqing, Prefect of Ningbo. According to records, during the Jiajing reign of the Ming dynasty, Qian Xiuyun, a smart and talented lady who loved reading, entrusted Prefect Qiu as a matchmaker and married Fan Bangzhu, a *xiucai* (a scholar who had passed the county-level imperial examination) and a descendant of Fan Qin, in the hope that she could get access to the books in Tianyi Pavilion. To her disappointment, however, after marriage, still she was not allowed to go upstairs to read, because the clan regulations forbade women to enter the library. Eventually she died with regret.

# II. Feng Mengzhuan and Fufu Room

"To withdraw from the society in the countryside, not for fame but for knowledge" is the lifelong pursuit of Feng Mengzhuan, a well-known modern book collector in Ningbo. The name of his private library "Fufu Room" (*lit.* the room of curled-up insteps) originated from the sentence "The cunning rabbit curled up to hide beside the insteps" in "Ode to Lingguang Hall in the State of Lu", with a collection of 100,000 volumes. Feng Mengzhuan (1886–1962), whose real name was Zhenqun, with courtesy names Mengzhuan and Manru, and assumed names Fufu Jushi, Chenghuazi and Miaoyouzi. In his later years, he named himself Lonely Old Man. Originally living in Cixi, his ancestors moved to the bank of Shuifu Bridge in Ningbo City. At the age of 17, he passed the imperial examination in the Renying year of the Guangxu reign (1902) and became a student of Ningbo government-owned school. In 1932, he served as the chairman of Yin County Literature Committee, and was engaged in honoring the sages and protecting cultural relics. Feng Mengzhuan dedicated his life to the collection and organization of local documents. He once led the renovation of Tianyi Pavilion, compiled the bibliography of the library collection, initiated the renovation of Baiyun Manor and commissioned the building of "Forest of Steles in Mingzhou". Even during the Chinese People's War of Resistance against Japanese Aggression, he built an air raid shelter in the *tianjing* courtyard of the Fufu Room for the book collection. In the case of air attack by Japanese air force, he never left the books for one moment, so that the collection could be preserved. He reached a conclusion from the history of Tianyi Pavilion that "books are difficult to gather but easy to disperse, and it is even more difficult for the descendants to preserve them forever". In his later years, he considered how to properly handle the collected books, his life's hard work, so as to prevent the collection from being separated or lost in other countries. In 1962, before his death, he told his descendants that "I have had a collection of hundreds of thousands of volumes, and now I'm going to contribute them to the country". He donated them all to the country. His noble character is self-evident.

Built in the Qing dynasty, Fufu Room faces east and is located next to Xiaowen Street, Haishu District, Ningbo. The whole building is a timber structure with five bays, two lanes and three wing rooms, which is simple, exquisite and well preserved. Entering the gate, there is a *tianjing* courtyard. On the right is a orange tree, whose branches cover a small half of the courtyard like a huge umbrella and whose roots are as thick as a bowl. It is said that though *daidai* oranges are not delicious, they can be used for medicine. It is assumed that Feng Mengzhuan planted this tree, hoping that the collection of books can be handed down to subsequent generations, because "*daidai*" in Chinese means generation after generation. Next to the orange tree grows an Osmanthus tree. In the courtyard are also some plantains with huge leaves, displaying a certain charm in ancient paintings. There is a hemispherical fortress of the height of half a man under the trees, leading to an air raid shelter. The couplets on the columns opposite to the fortress is the famed calligrapher Sha Menghai's handwriting: "A houseful of ancient books is preserved for present-day use; the contemporary knowledgeable man is immortal with the books." Mr. Feng Mengzhuan's life stories are presented in the main hall and the side rooms.

# III. Xu Shidong and Shuibei Pavilion

Xu Shidong, the owner of Shuibei Pavilion, was a famous book collector in Eastern Zhejiang in the Qing dynasty. Since he lived in the war years after the Opium War and suffered two wars, many of his works have been lost, and his collected books have been destroyed many times. However, he did not give up in finding them back. His story of book collecting from Yanyu Building to Chengxi Cottage, and then from Chengxi Cottage to Shuibei Pavilion was rare in the history of Chinese private libraries.

Xu Shidong, with the courtesy names Dingyu and Tongshu, and the assumed name Liuquan, is a native of Yin County. He was born in the nineteenth year of the Jiaqing reign of the Qing dynasty (1814) and died in the twelfth year of the Tongzhi reign (1873). In the twenty-sixth year of the Daoguang reign (1846), as a *juren* after passing the provincial imperial examination, he was appointed as *Zhongshu* (secretary, in charge of drafting, recording, translating and hand-copying official documents) of the Grand Secretariat. Since the seventh year of the Tongzhi reign (1868), he had been in charge of the compilation of *The Annals of Yin County* until his death in the twelfth year of the Tongzhi reign. He was fond of collecting books all his life. He copied and collated books by hand, often working all night long. He wrote more than 30 kinds of books and did the collation of *Six Annuals of Siming in the Song and Yuan Dynasties*. He examined and corrected mistakes in these books. These books are more valuable editions. His attainments were great enough to surpass his predecessors.

Xu Shidong originally lived in Yanyuzhou (now No. 48, Gongqing Road, Ningbo City) on the west bank of the Moon Lake, so his library there was called Yanyu Building, which was built in the Daoguang reign. The library comprises two sections of five-bay buildings. As the most famed cultural scenic spot among the Ten Islets of the Moon Lake in Ningbo, Yanyu Building had up to about 100,000 volumes of books in collection, mostly from Zheng Xing's Erlao Pavilion in Cixi, while some were from other ancient book collectors including Fan E'ting, Qiu Xuemin, and Hu Luting. In the last year of the Xianfeng reign (1861), many books in Yanyu Building were stolen in the war, and some were even burned for ignition by ignorant people. Thus, few books were left. In the first year of the Tongzhi reign (1862) when Xu Shidong moved to the Xicheng Cottage outside the West Gate of Ningbo, now No. 2,

Hengliu Lane, in the west of Ningbo City, he reorganized the old editions and found about 40,000 volumes of lost books. Unexpectedly, on the twenty-ninth day of the 11th lunar month in the second year of the Tongzhi reign (1863), a fire broke out and the books was destroyed. In the sixth lunar month of the third year of the Tongzhi reign (1864), he rebuilt a house at the former site of Xicheng Cottage in north of water and named it "Shuibei Pavilion" (*lit.* the pavilion in the north of water). After several years of collection, the books obtained reached 30 shelves altogether, including 798 kinds, 9,815 *ce* and more than 44,000 volumes. He compiled *Jing* (classics), *Shi* (history), *Zi* (philosophy), *Ji* (literature) and Series, gradually restoring the previous prosperity. In the eighth year of the Tongzhi reign (1869), the Bureau of Yin County Annals moved to Shuibei Pavilion. Xu Shidong presented his collections for people's need. In the third year of the Xuantong reign (1911), less than 40 years after Xu Shidong's demise, all his books were sold to Shanghai booksellers, with only a small number of them obtained by modern Ningbo collectors and Tianyi Pavilion.

Shuibei Pavilion used to be a two-story building at No. 2, Hengliu Lane, Ningbo. The building is basically intact, but has long been transformed into a folk house; the cooking smoke around the house brought about fire risk. In September 1994, due to the expansion of the road, the building could no longer be kept at the original site. As a result, it was moved to the south garden of Tianyi Pavilion Museum and restored to its original appearance for protection.

# IV. Lu Zhi and Baojing Building

Baojing Building was a famed private library in Eastern Zhejiang in the Qing dynasty. In those days, its richness in book collection was comparable to

Fan's Tianyi Pavilion and Zheng's Erlao Pavilion. Baojing Building and the Baojing Hall owned by Lu Wenchao in Western Zhejiang were known as the "East and West Baojing Buildings".

Han Yu, a poet of the Tang dynasty, wrote in a poem for Lu Tong, "Three commentaries on the *Spring and Autumn Annals* were put away unheeded, while the remaining classics held in arms were thoroughly studied in solitude." This is where the inspiration of the name Baojing (*lit.* hold the classics in the arm) Building came from one thousand years later, indicating the owner's determination to study the classics.

Baojing Building belonged to Lu Zhi, whose courtesy names are Danbi and Qingya. He was born on the third day of the fourth lunar month in the third year of the Yongzheng reign of the Qing dynasty (1725), and died on the eleventh day of the tenth lunar month in the fifty-ninth year of the Qianlong reign (1794) when he was 70 years old. The following is recorded in his household biography. "When he was young, he was extraordinarily intelligent. He learned from a Provincial Graduate named Guo Yonglin. At the age of 19, he was known to a Mr. Peng in Changzhou, a Provincial Education Commissioner, and thus became a government student of Yin County. He soon acquired a widespread reputation as a straight high achiever in every examination. In the thirteenth year of the Qianlong reign (1748), Mr. Yu in Jintan District (in Jiangsu) offered to enroll more students. In the nineteenth year of the Qianlong reign (1754), the county was suffering a severe famine, and Lu Zhi sent the grains there to help with the relief. His deeds were reported by senior officials to the emperor, and therefore he was awarded by the emperor's decree to become a tribute student (one of the top performers designated under the Directorate of Education for further study and civil service). Later, he applied for positions in the Provincial Administration Commission but failed repeatedly. He then was given the position of Secretary of the Central Drafting Office, according to the rules. Before the appointment date, however, he lost the sight of both eyes, so he was determined to give up."

Lu Zhi liked to collect books all his life. Whenever he came across a rare book, he would not hesitate to buy it at any cost. If his friends had any rare book, he would tactfully borrow and copy it. "He worked so hard at proofreading the books day and night that he often skipped his sleep and food often. Over more than 30 years, he had obtained tens of thousands of volumes of books." After he got blind, he asked his disciples to read books to him. "As it was too boring for him to just listen, he joined in by playing the ocarina."

Lu Zhi compiled the bibliography in person, and classified his collection of books into four categories: *Jing* (classics), *Shi* (history), *Zi* (philosophy), and *Ji* (literature), and then arranged them in order. In the thirty-ninth year of the Qianlong reign (1774), the Qing government granted Fan Qin's Tianyi Pavilion 10,000 volumes of *Complete Collection of Pictures and Books of Old and Modern Times*, which caused a sensation in Ningbo. Lu Zhi's "family has been wealthy for generations... He determined to compete with Fan's Tianyi Pavilion". He felt so sorry that he has not obtained such an imperial court edition that "he sent his servants to search the market for the books even at the cost of his bankruptcy. When the books arrived, he dressed himself up to welcome them at the gate. Obviously, he was terribly addicted to book collection".

The two-story library was built in the east of Lu Zhi's Residence, imitating Tianyi Pavilion in style, with six bays facing south. There is a hall in the middle downstairs and stairs leading to the second floor in a west room. The stairs are slightly different from those in Tianyi Pavilion in that they are laid transversely. Books are stored upstairs, separated by bookcases. As is seen from the sketch of bookcases arrangement in Baojing Building in the late Qing dynasty, two large single-sided bookcases are placed on both sides of the wall. In the middle are ten large bookcases in five rows, which can be opened in front and back. In the south are ten small bookcases. In front of the library were rockeries and a square pond surrounded by bamboos and trees. In the sixth lunar month of the third year of the Xuantong reign (1911), Feng Mengzhuan went upstairs to read books and wrote about it. At that time, Baojing Building "was decorated in a simple and plain style, without a trace of vulgarity. With towering trees around, the windows, the walls and the balustrades were all soaked in the green".

Lu Zhi held Tianyi Pavilion in high regard. Not only did his library copy it in style, but he also learned from it in the management of books and set the rules that the books be jointly owned and managed by future generations. Its strict management system was implemented until the end of the Qing dynasty. After Lu Zhi passed away, the library was closed.

Baojing Building was originally located in the southeast corner of Ningbo City, now No. 18 of Junzi Street, and the side gate is at No. 1 of Shiban Lane. The library was solidly built, and it still maintains its simple style through more than 200 years. The water ripple decorations on the ceiling and eave rafters of the central hall are clearly visible. The front windows upstairs are a row of *minghe* (*lit.* crane) windows, with six for each bay. In the back are two windows for each side bays. The upper half of the wooden windows are of grid patterns, each grid with a shell embedded in it, and several of windows still remain the same. In the middle hall downstairs, the original plaque inscribed with " 抱经楼 " written by the renowned scholar Ruan Yuan was destroyed later. The pond in front of the building had been filled up, and a small house had been built to accommodate people, with no restrictions on fire or candles. In 1995, due to the reconstruction of the old city, Baojing Building was demolished and the wooden structure is now kept in Tianyi Pavilion Museum. In 1998, the construction of Baojing Hall was completed to commemorate Lu Zhi.

# V. Sun Jiagui and Woji Hut

To the south of Tianfeng Pagoda, at the intersection of Kaiming Street and South Jiefang Road, there is an alley called Taqian Street. No. 23–24 of the alley is "Woji Hut". It was once a famous private library in Ningbo, and the birthplace of Sun Chuanzhe, the designer of the first stamp in the new China.

Entering the door at No. 23 in the lane, we can see a six-bay two-story wooden house facing west and a three-bay wooden house facing south, forming a small courtyard of dozens of square meters. The building belongs to a person whose surname is Sun. According to research, the Sun family was originally from Sunjiajing Village in Henghe, Cixi. In the late Qing dynasty, they moved to Ningbo and purchased the house from the eldest son of the Yao family next door and settled down here.

Sun Jiagui (1879–1946), with the courtesy name Xiangxiong, showed no interest in any official career all his life, while used to be a private school teacher and a store clerk. He was so fond of ancient classics that he was especially diligent in seeking old books, and the collection gradually became rich. He turned the middle room into a place for his collection of books. Because of its small size, he referred to it as "Woji Hut" (*lit.* a place where a snail settles in). Most of the books in Woji Hut were purchased in the period from 1915 to 1930 when the book collectors were eager to sell their collected books, due to the wars between warlords and the social unrest. Besides, Sun Jiagui was willing to pay a high price, so he obtained many good books. Among them, there were many rare books from the famous private libraries, such as Tianyi Pavilion, Baojing Building, Yanyu Building, and so on. Sun Jiagui hired people to copy rare books up to 20,000 volumes, especially the stencil tissue paper edition books of the Ming dynasty. The collection of poetry and essays is a major feature of Wojilu Library. Many celebrities in Ningbo have paid their visit to Woji Hut.

Sun Dingguan (1903–1985), the eldest son of the Sun family, influenced by his father, loved ancient books, calligraphy and paintings when he was young. After inheriting the books in Woji Hut, he cherished them so much that he often sorted and mended them by himself, and continued to buy more. In 1979, Sun Dingguan donated 954 books and 66 works of calligraphy and painting to the library of Tianyi Pavilion. In 1987, Sun Shile, the eldest grandson, gave away all the remaining books to Ningbo University to fulfill the last wishes of his grandfather and father.

# Chapter Eleven

# Ancient Stages in Ningbo: The Carrier of Opera Culture

The taverns are crowded with drinkers, like markets in the morning;

Opera actors and actresses give their show in the temple in the evening.

Through field paths on a yellow calf, a drunken teacher

Went to the countryside to watch the opera.

This is a poem by Lu You, a well-known poet of the Southern Song dynasty, describing the scenes of opera shows in the rural area, which reflects the important role of opera in the old rural life.

Ningbo operas play a vital part in southern operas. Its main representative is *The Story of Pipa* written by Gao Ming, a dramatist of Yin County in the late Yuan and early Ming periods. It was important in the history of Chinese opera and is known as the "ancestor of opera". Ningbo is home to Yongju opera (also Ningbo opera), which originated in the rural areas of Eastern Zhejiang and became widely spread across Ningbo, Zhoushan, Taizhou and Shanghai. Yuyao opera has a longer history. Previously as one of the four major tunes of southern operas, now it still survives as a local opera. Ningbo is also the second hometown of Shaoxing Opera, which originated in Shengxian County, Zhejiang

Province in the late Qing dynasty and developed greatly in Ningbo in the late Qing and early Republican period. Due to the prosperity and development of traditional operas, the stage, as a performance place, came into being.

According to relevant records, the earliest stage appeared in the Tang dynasty. There are pictures of stages in the murals of the Mogao Grottoes in Dunhuang, showing that the stages, without top cover, are made of earth or wood. The earliest record of the opera stage is on the "Three Steles of Operas". Among them, the "Monument to the Temple of Earth Goddess" built in the fourth year of the Tianxi reign of the Northern Song dynasty (1020) mentions a dancing pavilion, which is considered to be the earliest temple stage in China. Unfortunately, no physical stages of the Tang or Song dynasties can be found.

The emergence of opera stages made of the earth marked the origin of Chinese opera. Through the primitive songs and dances in the Tang and Song dynasties to the ripe age of operas in the Yuan dynasty, the opera witnessed prosperity in the Ming and Qing dynasties. During the Ming and Qing dynasties, opera pavilions with upturned eaves and protruding corners appeared. Buildings, pavilions and even gardens were built for opera performance and appreciation inside temples, guild halls and ancestral halls, along the riversides, and in the streets. The construction technology of stage buildings became increasingly sophisticated, the auditorium facilities continued to be improved, and the stage developed from one stage in one temple (or ancestral hall) to two or even three stages.

Ancient stages in China generally have the following architectural features.

In terms of layout features, first of all, the idea of deification is highlighted. Temporary stages, such as *caotai* stage (the stage that is made temporarily with anything possible, like several tables put together) and *chuantai* stage (the stage that is assembled temporarily with separate wooden parts), are not limited by the environment and are highly random, while the street stages and the road stages have a simple relationship with the surrounding environment. In the overall design of ancestral halls and temples, the stage is arranged as a main building to be situated on the vertical axis in a main position. From front to back, there are the entrance hall, the stage, the main hall, and the rear hall. On the left and right are wing rooms, and surrounding the stage a quadrangle-style building complex is built. The orientation of the stage is opposite to that of other buildings. It faces the main hall, and the god tablets and the ancestral memorial tablets were hosted. In the old days, acting was performed for the sake of gods and ancestors because people saw gods as the savior who could bring the world peace, prosperity and good luck, prevent disasters, and avoid risks. Therefore, on the birthdays of gods or ancestors, the performance would last for three days and nights, and grand temple fairs would be held to show their celebration and respect for gods. The stage board of City God Temple in Ningbo can be disassembled for easy moving. The Temple seated statues and walking statues, and the walking ones are smaller in size and can be moved. Every festival when it was time for the City God to go out for an inspection of people's life, people carrying the walking statues would go through the main gate and past the stage. Therefore, the stage board must be removed and reinstalled after returning to the temple, which also reflects the idea of deification.

Secondly, orderliness is highlighted. The main hall is five to ten meters away from the stage, and the backstage is connected to the main gate. The second floor of the main gate is higher than the stage deck but it has stairs leading to the stage, which is therefore called *daogua* building. The second floor of the main gate is used for prop storage, lounge and make-up room for actors and actresses during performance. The performance area of the stage is separated from the back stage by a screen standing in the shape of Chinese character " 八 ". On the screen, Three Gods "Fu, Lu and Shou" (Gods of Happiness, Fortune, and Longevity) are painted. Behind the screen is the so-called "backstage", where the seats were set specially for the band. Some stages are equipped with an extra stage, which can be assembled and disassembled any time. The stage temporarily set up at the east end is mainly used for the band's playing and is disassembled after the performance. There are wing rooms to the east and the west of the stage. The second floor is the audience area specially for women and children. In the old days, men and women were not allowed to have any physical contact, and it was stipulated that men should not go upstairs. The stage is surrounded by the entrance hall, the main hall and the wing rooms on both sides to form a courtyard shaped like Chinese character " 凹 ", which serves as the audience area, or the ground where the audience sit and watch the opera. In addition to accommodating the viewers, the courtyard also has the following three functions: lighting, ventilation and drainage. Since there are no windows on the outer walls of the ancestral halls and temples, lighting, ventilation and drainage of the roof water all rely on the courtyard. A few stages are connected with the main hall to ensure that the audience area will not be affected by wind and rain.

Third, in terms of space occupation, the stage is designed to guarantee the normal functioning and the impressiveness of the whole building. The ground is of the same level as the main gate, five to seven steps lower than the main hall. This makes the main hall look tall and magnificent. The height of the stage deck is generally 1.6 meters to 2 meters, which guarantees a good view for the audience, whether they are in the courtyard, the main hall or the wing rooms. The top of the main gate, the roof of the stage and that of the main hall increases in height, forming an invisible step, which means going up step by step.

In terms of structural features, the stage integrates the foundation of the multi-story building, the beam frame of the temple and the roof of the pavilion. The stage can generally be divided into three parts, namely the foundation, the beam frame and the roof.

**1. The stage foundation.** The foundations are built according to local conditions. In terms of *maitou* (the corner part of building foundation) stone is applied for the colunmn foundation, and rammed earth is used for other places. The *taiming* part (the surrounding edge surface and outside walls of the building foundation) of the foundation include a square slab twice the size of the rock pier and a column base placed on it. There are many flower pattern decorations on the column base, mostly the Chinese character " 亞 " ( 亚 ). Some colunmn bases are drum-shaped, girdle-shaped, flower-basket-shaped, square-shaped. Most of them are plain, while some others are carved with various kinds of patterns of *ruyi*, Chinese character " 回 " , animal faces, and so on. The durable stone column bases were used because of the humid and rainy climate in Southern China. The edges around the stage are laid with rectangular flat stones, 10–15 centimeters above the ground. The stage deck is mostly made of tabia, and a few are paved with stone slabs.

**2. The column frame.** All columns are stone columns, and a few are wooden columns. The upper end is round, while the lower end is square. Columns are 4.5–5 meters high, with a diameter between 25–30 centimeters. Most stages use ten square auxiliary posts to support the stage fence. The two stone columns at the front of the stage are 40 centimeters high from the deck, decorated with a pair of carved lion heads atop, facing each other, which means to guard against evil spirits. The upper ends of standing columns are made into cross-shaped mortises, which are used to lock the upper architraves. Two upper architraves are locked by an upper and a lower mortise into the column top to support the lower architrave above it. Between the upper and the lower architraves is bracket set. The *douqi* is generally the seven-*cai* and three-*ang* bracket set, with intercolumnar bracket sets and column-top bracket sets *dougong* placed together on the side. Between the intercolumnar bracket sets is decorated with *jiatang* board in between, and the lower architrave is supported by an eave purlin. Due to the necessary use of circular caisson ceiling, its tapering design is somewhat different from the gable-and-hip roofs of other palace buildings. At the outer one-third of the eave purlin are placed four upward sloping *caibujin*, which forms a triangle with two eave purlins in different directions, and then put five-purlin beams or three-purlin beams on it, with dwarf columns and *chashou* (inverted V-shaped brace) on it to support the ridge purlins and then to form the frame of a gable-and-hip roof. The four roof corners adopt the method of *faqiang* (upturning). One end of the *laoqiang* is fixed on the head of three-purlin beams, and the other end is placed on the column-top bracket set of the front column, extending downward and outward. The *laoqiang* (the sloping piece of wood ready for the protruding corner of roof) forms a 45-degree angle with the *nenqiang* (the upturning piece of wood to make the protruding corner of roof), which is covered with triangular wood, rhombic wood and shoulder pole wood, arc-shaped and upturned to form a gracefully stretched curved roof corner.

**3. The roof.** Four purlins are used, namely the ridge purlin, intermediate purlin, eave purlin and overhanging eave purlin. The roof between the eave purlin and the intermediate purlin rises gently, but the higher it goes up, the larger amount of *jushu* (the vertical rise of the roof) is, and the steeper the eave slope is. Generally, the vertical distance from the eave purlin to the lower intermediate purlin 50% of the horizontal distance, that between the two intermediate purlins is 70% to 80% of the horizontal distance, and that between the upper intermediate purlin and the ridge purlin is 90% or even more than 100% of the horizontal distance. As the saying goes: "The stage is beautiful, but its roof is difficult to build." It is very hard to stand on such a steep slope. So the construction was difficult, as the roof might slide any time. Whenever it gets loose, the rain leaks in. The stage roof is built into a curving shape, downward in the middle and upward at the ends, by adopting different horizontal distances between different purlins, which ensures not only the drainage of the roof, but also the three-dimensional aesthetic effect. In order to produce far-reaching eaves without making the eave ends look too low and thus affecting the lighting, flying rafters are used to slightly lift up the eaves. To prevent the tiles from sliding, the rafters are covered with roof boarding or brick sheathing. A thick adhesive layer of *huibei* (an adhesive substance to stick tiles together) is applied to prevent roof tiles from falling. The roof is covered with small black tiles or barrel tiles, which are laid in an extremely dense manner, with eight covered and two exposed over the *jibu*, and seven covered and three exposed over the *jinbu* and the *yanbu*.

There are many ancient stages in Ningbo, and hundreds of them are on record. Among them are stages in ancestral halls, such as the stage of Qin's

Branch Ancestral Hall to the west of Moon Lake, Ou's Ancestral Hall stage in Qiangtou Town of Xiangshan County, and the Dunshan Hall stage and the Xiaoyou Hall stage in Qingtan Village of Ninghai County. There are also guild hall stages, such as those in Qing'an Guild Hall and Anlan Guild Hall in former Jiangdong District. There are also sacrificial temple stages, such as those in City God Temples of Ningbo City, Ninghai County, and Shipu, Xiangshan. Among them, the stages of City God Temple in Shipu, Xiangshan is rare in Ningbo in that there are two stages in one temple; other examples are the stage of the Temple of Guan Yu at the Moon Lake, the stage of Xiansheng Temple in Yuyao, the stage of Xiaowang Temple in Fenghua, the stage of Huanggong Temple and the stage of Zhongying Temple in Yinzhou. A fourth type is the stages along the riverside or on the street, which are not many, such as the riverside stage on the opposite bank of Huaguoyuan Temple at the Moon Lake, the street stage in Huanggulin, Yinzhou, the street stage in Juexi, Xiangshan, and so on.

The earliest extant stage in Ningbo is the Xiansheng Temple stage in Luting Township, Yuyao. According to the records, it was first built in the Southern Song dynasty and moved to this site during the Zhizhi reign of the Yuan dynasty (1321–1323). The existing buildings were rebuilt during the Kangxi reign of the Qing dynasty. The square stage is 4.7 meters wide and about 6 meters high, with a gable-and-hip roof and the upturned eaves and protruding corners. The bracket set on the top of the stage spirals upward and shrinks inward to form a spiral caisson ceiling, commonly known as the *jilongding* (the type of spiral caisson ceiling that looks like the top of a chicken cage). There is also a large bronze mirror in the center of the top, surrounded by eight dragon heads and flower baskets. The bronze mirror and caisson ceiling in the dome not only symbolize the meaning of fire prevention and disaster avoidance; according to scientific measurement, the four walls of the ceiling create resonance with the singing voice of the performers and make it more sonorous.

The most beautiful existing stage in Ningbo is the one of Qin's Branch

Ancestral Hall in Tianyi Pavilion Museum. It faces north, with several lifelike dramatic figures standing on the top of the gable-and-hip roof. The two stone columns in the front of the stage were replaced by steel pipes with a diameter of about 12 centimeters for the audience to have a better view of the stage. The stage is divided into a front stage and a back stage, separated by a wooden screen wall with doors on both sides, which are the so-called "*chujiang*" (outgoing of a general) and "*ruxiang*" (incoming of a minister), respectively. "*Chujiang*" is the exit through which an actor walks onto the stage, while "*ruxiang*" is the entrance through which an actor walks off the stage. In the old days, they were also collectively referred to as "*gumen*" (ancient gate), which means that what was staged there was all about ancient stories. The backstage is the opera room close to the entrance hall. It is also the place where the sound effects are made during the performance, like barks of dogs, sound of rain, sound of thunder and so on. This is also the waiting place for the actors and actresses. On the left side of opera room is the dressing room, and on the right is the room for the setting, props and clothes. In front of the stage is a *tianjing* courtyard. The stage faces the main hall, and there are audience seats on the second floor on its left and right. In the past, whenever there was an opera performance, the villagers, old and young, would gather around the stage. The music sounds were like a magnetic field, attracting people to the stage and bringing them a simple joy.

The time-honored stages are not only a vivid history recording the development of opera, but also a collection of couplets. Generally, there are couplets inscribed by ancient literati preserved on the two columns in the front of the ancient stage. For example, the couplets of the street stage at Huanggulin, Yinzhou District, read: "In the transportation hub the opera entertains the passers-by; beside the great port the melodies flow into the water nearby." The couplets on the stage of the Temple of Guanyu of the Moon Lake read: "People are watching the opera at the Moon Lake like the jade pot; stories are played on the stage like the history itself." Those on the stage of Dunshan Hall in Qingtan Village, Ninghai County, read: "The fictional stories in the opera are played to reveal the reality; the ancient characters in the plays are meant to remind you of history." Another pair of its couplets read: "People are depicted with sound and makeup; the past and present history is told without words." Those on the stage of Shuangzhi Temple in this village read: "A song of spring awakens ancient and modern dreams; two kinds of faces show loyalty and evil." The thought-provoking contents of the couplets on the ancient stages are either indicative of the environment, or alluding to the present via the past anecdotes.

In recent years, Ningbo cultural relics departments at all levels have done a lot of work to protect ancient stages. They conducted an inch-by-inch survey, re-evaluated the well-preserved ancient stages, selected and successively announced dozens of them, such as Juexi Street Stage Pavilion in Xiangshan, as historical and cultural sites protected at different levels or candidates for the titles. In May 2006, the ten most representative and artistic ancient stages in Ninghai County were announced as the sixth batch of Major Historical and Cultural Sites Protected at the National Level.

Now, let's go to these ten ancient stages one by one and appreciate their profound cultural heritage by taking a look at their exquisite structure and gorgeous decoration.

### 1. The Ancient Stage of Chongxing Temple

The ancient stage of Chongxing Temple is located between Shijia Village and Houxi Village in Xidian Town and is shared by the two villages. Both as the descendants of Shi Xianwen, Grand Master for Forthright Service in the Qiandao reign of the Song dynasty (1165–1173), the two villages belong to the same clan and share the same surname "Shi". By the middle of the Kangxi reign, Shi Chengwo (1643–1722) built Chongxing Temple. Shi Yuntai moved the temple to the left of Shi's ancestral hall in the twenty-first year of the Daoguang reign (1841), and built the stage and the triple caisson ceilings at the same time.

## 2. The Ancient Stage of Hu's Ancestral Hall in Aohu Village

The ancient stage of Hu's Ancestral Hall is located in Aohu Village, Meilin Street. In the second year of the Jiaqing reign of the Qing dynasty (1797), Hu Yuanshi, a *yixiangsheng* (another name for *xiucai*, a scholar qualified for the local level imperial examinations), built Hu's Ancestral Hall, called "Jiqing Hall". At that time, the front hall was relatively simple, just a three-bay, single-story building. In the fourth year of the Xianfeng reign (1854), Hu Yinjie took the charge in rebuilding the front hall into a two-story building, funded by the families of the clan. The building was built by means of *pizuozuo* (splitting and building, i.e., two groups of craftsmen work for the construction of the same building from opposite directions and compete with each other, so that the style of building on the two sides is slightly different but still in harmony). In the 1920s, the stage and the connected corridor (commonly known as the I-shaped house) were reconstructed and triple caisson ceilings were added to the stage.

## 3. The Ancient Stage of Wei's Ancestral Hall in Xiapu

The ancient stage of Wei's Ancestral Hall in Xiapu is located between Houzhou Village and Xiayang Village in Qiangjiao Town. The ancestral hall is shared by the two villages, which are collectively called Xiapu. The descendants of the Wei family built a three-bay hall in the eighth year of the Kangxi reign of the Qing dynasty (1669), and expanded it into five bays in the Daoguang reign (1890). In the sixteenth year of the Guangxu reign (1890), the *yimen* gate, the stage and wing rooms were built by the method of "*pinzuozuo*" on the East-West axis, which accounts for their distinct styles on the two sides.

## 4. The Ancient Stage of Pan's Ancestral Hall in Panjia'ao Village

The ancient stage of Pan's Ancestral Hall is located in Panjia'ao Village, Qiaotouhu Sub-district. It was proposed by Pan Jiaxing, Pan Jiasi and Pan Jiayu in the Qianlong reign of the Qing dynasty (1784) to build a three-bay ancestral hall with a stage. In the Jiaqing reign (1810), the patriarch Pan Jiaqi took charge of the construction of the two-story, five-bay front hall. In 1922), the patriarch Pan Dapin together with the person in charge did a major overhaul of the stage and the two wing rooms. Pan's Ancestral Hall was also built by the method of "*pizuozuo*".

### 5. The Ancient Stage of Shuangzhi Temple

The ancient stage of Shuangzhi Temple is located in Qingtan Village, Shenzhen Town. Shuangzhi Temple has always been the temple of Taoism gods in Li'ao. It was initially built by Zhang Shishang and Zhang Tingyu in the Zhengde reign of the Ming dynasty (1506–1521), and later it was abandoned and renovated time and again. In 1933, it was rebuilt with funds raised by Zhang's family, Zhu's family and Kong's family from six natural villages including Qingtan Village.

### 6. The Ancient Stage of City God Temple

The ancient stage of City God Temple is located on South Taoyuan Road, Yuelong Sub-district. The City God Temple was first built in the first year of the Yongchang reign of the Tang dynasty (689) and rebuilt in the first year of the Longxing reign of the Southern Song dynasty (1163). In 1935, a large-scale maintenance was carried out on the City God Temple. The four square columns in the front of the stage that blocked the audience's view were removed and replaced with two iron columns. The extant buildings, including the *yimen* gate, the stage and two wing rooms, were all built at that time. In 2002, a thorough maintenance was carried out again.

### 7. The Ancient Stage of Chen's Ancestral Hall in Longgong Village

The ancient stage of Chen's Ancestral Hall is located at the entrance of Longgong Village, Shenzhen Town. Chen's Ancestral Hall was built in the early Qing dynasty. It has a beautiful environment, with Longxi River in the south and Shishan Mountain in the north. From the south to the north in the complex are the screen wall, the front *tianjing* courtyard, *yimen* gate, the middle *tianjing* courtyard, the middle hall, the stage, the rear *tianjing* courtyard and the main hall.

## 8. The Ancient Stage of Yu's Ancestral Hall in Ma'ao Village

The ancient stage of Yu's Ancestral Hall is located in Ma'ao Village, Shenzhen Town, and was built in the eighth year of the Wanli reign of the Ming dynasty (1580). In the fifth year of the Shunzhi reign in early Qing (1648), the ancestral hall was burned down by the Qing army due to the "Baitouweng" (Chinese bulbul) uprising led by Yu Shusu. It was rebuilt on its original site in the nineteenth year of the Kangxi reign (1680). A fire broke out in the second year of the Xuantong reign (1910). In 1912, Yu Mincheng was recommended to raise funds and to rebuild it. The stage was then erected.

The stage has a gable-and-hip roof, without ridge decoration, and the caisson ceiling is made into a spiral shape with the combination of specially-shaped bracket set and *ang*.

## 9. The Ancient Stage of Hu's Ancestral Hall in Dacai Village

The ancient stage of Hu's Ancestral Hall is located in Dacai Village, Shenzhen Town. The ancestral hall was first built in the Southern Song dynasty, and has been abandoned and restored several times since then. Its current scale was built in the late Qing dynasty, and the caisson ceiling of the stage was built at the same time. The ancestral hall faces north. Along the central axis, there are the screen wall, the front *tianjing* courtyard, *yimen* gate, the stage, the connected corridor and the main hall.

## 10. The Ancient Stage of Lin's Ancestral Hall in Jiajueke Village

The ancient stage of Lin's Ancestral Hall is located in Jiajueke Village, Qiangjiao Town. The Lin family moved here from Renheli, Hangzhou during the Jiading reign of the Southern Song dynasty (1208–1224). The initial time of constructing the ancestral hall remains unknown, but the existing scale was built in the late Qing dynasty.

Part III

# Modern Architecture in Ningbo

# Chapter Twelve

# Ningbo Commercial Guild Halls with Unique Styles

The guild hall is a type of urban public building in China, which refers to the halls jointly built by fellow countrymen from the same places for the gathering or temporary accommodation of fellow countrymen of the same trade. So far, the earliest guild hall in China is the Wuhu Guild Hall in Beijing built in the Yongle reign of the Ming dynasty (1403–1424). The guild halls were flourishing a century later in the Jiajing and Wanli reigns, and reached their peak in the Qing dynasty. The commercial guild halls in the Ming and Qing dynasties experienced the transition from *huiguan* (guild hall) to *gongsuo* (guild office) and then to *shanghui* (business association), which thrived until the late Qing dynasty and the period of the Republic of China.

Guild halls generally include countrymen guild halls and industry guild halls. The former was a place for fellow countrymen coming from the same place and living in the same city to get together, keep in contact and stay temporarily. The latter was the place where a guild of commerce and a guild of handicraft held meetings and consultations and did businesses. The architecture of the countrymen guild hall is roughly similar to that of large-scale residential buildings, some of which are reconstructed from large-scale residential buildings. However, the industry guild hall has a different style. Although its general layout is similar to residential buildings, it pays more attention to decoration by using complicated carvings and gold decorations. Architecture is a physical manifestation of culture. The guild hall displays its functions and values in politics, economy, religion, culture, art and other aspects to varying degrees. Politically, the combination of guild halls and feudal forces has played a certain role in protecting the interests of businessmen under certain conditions. Religiously, the guild halls established by the inland commercial groups are always connected to the Temple of Guan Yu, while the guild halls established by the coastal commercial groups are always connected to the Queen of Heaven (Mazu) Temple. Economically, since the middle of the Ming dynasty, a large number of industrial and commercial guild halls emerged, and the guild hall system began to evolve from a township organization to an industrial or commercial organization, which played a part in boosting social, political and economic development.

The formation of Ningbo guild halls can be traced back to the Song dynasty. In the second year of the Shaoxi reign of the Southern Song dynasty (1191), Ningbo had a clear record of the guild hall at its initial stage, which is as follows. Shen Faxun, a Fujian shipping merchant, was trapped in an accident at sea, but was blessed by Mazu, Goddess of Sea. He then took the incense from Mazu Temple in Putian, Fujian back to his residence in Jiangxia, Ningbo, before donating his residence to the service of Mazu and having it consecrated as a temple. Hence the first Queen of Heaven (Mazu) Temple in the history of Eastern Zhejiang was built, and later it became the guild hall of the shipping industry group of Fujian, named Bamin Guild Hall ("Bamin" is Fujian; "ba", literally "eight", represents eight administrative divisions of Fujian Province; "Min" is the short form for Fujian Province).

During the Tianqi and Chongzhen reigns of the Ming dynasty (1621–1644), Ningbo herbal medicine merchants first settled in Beijing to expand the market and set up "Yin County Guild Hall", which marked the emergence of Ningbo guild halls in other places.

The Ningbo Commercial Group not only established commercial guild halls in Beijing, Tianjin, Shanghai, Nanjing and other metropolises, but also set up guild halls in Japan, Singapore and other countries in Asia in order to unite businessmen of the Ningbo Commercial Group overseas. The frequent gathering of and contact between commercial in Ningbo and those who come to Ningbo for business from all over the country created a splendid culture of commercial guild halls.

In the late Qing and early Republican period, there were multiple famous commercial guild halls in Ningbo, including the Bamin Guild Hall, also the Queen of Heaven (Mazu) Temple, of the Fujian Commercial Group on Jiangxia Street, the Banking Guild Hall at the entrance of Zhanchuan Street, Fujian Old Guild Hall mainly dealing in timber located in former Jiangdong District, Qing'an Guild Hall of the Ningbo Beihao (trade in Northern China) maritime merchants, Anlan Guild Hall of Ningbo Nanhao (trade in Southern China) maritime merchants, Lingnan Guild Hall of Guangdong Commercial Group, Lianshan Guild Hall of Shandong Commercial Group, and Xin'an Guild Hall of Huizhou Commercial Group. In addition, there are Minzhe ("Min" is short for Fujian and "Zhe" for Zhejiang) Guild Hall located at Zhaobao Mountain in Zhenhai, and Minguang ("Guang" is short for Guangdong) Guild Hall and Sanshan Guild Hall in Xiangshan.

With the passage of time and the changes of the city, there are not many commercial guild halls left in Ningbo, with only a few well-preserved ones on records, among which the best known are Qing'an Guild Hall, Anlan Guild Hall and the Banking Guild Hall.

Qing'an Guild Hall and Anlan Guild Hall, located at the Three-river Junction, are the only commercial guild hall complex integrating a palace [the Queen of Heaven (Mazu) Temple] and a guild hall in urban Ningbo.

In the Qing dynasty, those who engaged in southern trades were called "Nanhao", mainly dealing in Fujian timber, while those in northern trades are called "Beihao", mainly dealing in the special local products of Shandong.

Jiangdong, east of the Yongjiang River, which used to be desolate, then become a prosperous sailing port and a prime location to open stores, as many Nanhao and Beihao merchant ships were moored there. The guild hall of Nanhao was built in the Daoguang reign of the Qing dynasty (1821–1850) and was named "Anlan" (*lit.* to calm the waves of the sea), which means "praying to Mazu for a peaceful sea". The guild hall of Beihao was established in the third year of the Xianfeng reign of the Qing dynasty (1853), and was named "Qing'an" (*lit.* the celebration of peace), which means that "a peaceful sea deserves to be celebrated".

The two guild halls are well known not only because of their influence on commerce and Mazu culture, but also because of their ingenuity in architecture, which is second to none in Eastern Zhejiang.

The main building of Qing'an Guild Hall faces west, covering an area of about 3,900 square meters. On the gate of the guild hall, the surrounding internal and external walls, and beams are the brick carvings, stone carvings and red lacquered gilded wood carvings. Among them, the dragon-phoenix stone columns, the brick-carved palace gates, and the wooden caisson ceiling of the stage are recognized as the "Three Wonders" of Eastern Zhejiang carvings.

The "wonder" about the dragon-phoenix stone columns is represented by a pair of curled-up dragon stone columns and a pair of phoenix-peony stone columns in the main hall. The columns are more than 4 meters high, adopting the unique technology that combines high relief and openwork carving, which displays unique and rare carving craftsmanship in China coupled with the exquisite column bases. An awe-inspiring coiling dragon curls around the column with two bats flying up and down in the clouds. The two phoenix-peony stone columns are standing on both sides. The upper half is carved with a male phoenix and the lower half a female phoenix, both ready to fly. The walls next to the phoenix-peony stone columns are inlaid with two stone screens, which were delicately carved with the Ten Scenes of the West Lake in low relief; the relief of less than one centimeter deep is elegant and serene, in contrast to the wild and vibrant dragon and phoenix mentioned above.

The main gate is a medium-sized brick gatehouse. It is apparent that the owner did not want the guild hall to show off. The facade is a brick gatehouse. Its lintels are decorated with brick carvings of 14 stories and bracket sets with wood-like brick carvings. The stone carvings on the plinth are embossed with flower patterns, and the walls were built with smooth facing bricks. On the lintel is a vertical brick plaque in the shape of an imperial edict scroll, carved with the relief of double dragons playing with beads. In the middle are the Chinese characters of "Tianhou Palace" in relief. The brick carvings on both sides of the plaque include stories such as "Eight Immortals", "The Fisherman, the Woodcutter, the Farmer and the Scholar" (now reconstructed), as well as animal patterns, such as phoenix motif and that of lions rolling a ball.

There is usually a caisson ceiling over the stage. The caisson ceiling of the stage in Qing'an Guild Hall is a *jilongding,* the type of spiral caisson ceiling that looks like the top of chicken cage. It is made of hundreds of carved wooden boards tenoned together, the gold paint on them beautifully dazzling. At the four corners of the caisson ceiling are four bats symbolizing blessings, and on the wooden balustrade around the stage are carved several dragons playing with beads. The most amazing is the bracket sets, the *gualuo* (hanging fascia board) and the carved wooden boards around the top of the stage, which vividly demonstrate the exquisite workmanship of red-lacquered gilded wood carving in Ningbo. The carved wooden boards use relief carving, mainly narrating the stories of the Three Kingdoms, such as the "Three Heroes Fought Against Lü Bu". The three hanging fascia boards are carved with three pairs of dragons playing with beads and the phoenix playing with peonies using openwork carving. The bracket sets are carved into dragon heads and flying phoenixes. "*Chujiang*" and "*ruxiang*" are also made into the pattern of dragon, and the relief of six maids on the back wall of the stage are also vividly carved.

The architecture of Anlan Guild Hall is also remarkable in some ways. Its construction was funded by Ningbo Nanhao shipping merchants in the sixth year of the Daoguang reign of the Qing dynasty (1826), known as the "Nanhao Guild Hall". It is also a place for people of the same trade to meet and worship

Mazu. Its overall building faces west, comprising the main gate, the front stage, the main hall, the rear stage and the rear hall from west to east, covering an area of 1,700 square meters. Its architectural style is similar to that of Qing'an Guild Hall, with a gable of the *guanyindou* style (a style of gable resembling the hood of Guanyin Bodhisattva), tall and solemn. The stage is exquisite, the main hall is magnificent, and the round ridge roof, *queti*, and architraves all have delicate and magnificent red-lacquered gilded wood carvings. The purlins and beams of the central bay and two side bays are decorated with lifelike gold dragons and phoenixes. The brick carvings and the stone carvings as the decoration are exquisitely made with gorgeous patterns. There are two stages in the guild hall, one in the front and the other at the back, forming a unique pattern of two guild halls and four stages together with Qing'an Guild Hall, which is quite rare in China.

Anlan Guild Hall has a history of nearly 190 years. In 2000, Ningbo municipal government moved it to the south of Qing'an Guild Hall, making the two guild halls a perfect pair with each other.

Let's talk about the Banking Guild Hall. Ningbo's financial industry has always been dominated by *qianzhuang* or private banks. As documented in *The General Annals of Yin County* compiled in the period of the Republic of China, private banks are the financial hub of Ningbo. At its peak, there were 36 larger banks among them had funds of more than 60,000 yuan, and more than 30 smaller banks more than 10,000 yuan, with more than 160 banks in the urban area alone at its best. Ningbo people had a reputation for their diligence, intelligence and good management. During the Daoguang reign of the Qing dynasty (1821–1850), private banks in Ningbo initiated a system that allowed the money to be transferred between the private banks and transactions among all walks of life to be settled with the exchange of bills, making ready money no longer necessary in transactions. It marks the beginning of the bill exchange of modern financial industry in China, which was roughly at the same time as the founding of the first clearing house in London in 1833, but much earlier than New York, Paris, Osaka, Berlin and other cities. So far, in Banking Guild

Hall there are still stone steles describing the development of the financial industry of Ningbo and the process of building the guild hall.

Banking Guild Hall is located at No. 10 of Zhanchuan Street, not far from Dongmenkou in the urban area. It was recorded that in the third year of the Tongzhi reign of the Qing dynasty (1864), the organization of the banking industry had a *gongsuo* (guild office) in Bingjiang Temple near Jiangxia, which had been destroyed in war and later rebuilt by the banking industry. By 1923, considering "the original office was too small", the industry purchased a shipyard and a house of "Pingjinhui Foundation"[1] (now on Zhanchuan Street), and built a new guild hall, the current Banking Guild Hall, which was completed in 1926. It used to be a place for the meeting and trading of Ningbo's financial industry.

Covering an area of more than 1,500 square meters, the guild hall has a quiet and beautiful environment and convenient transportation. It is a brick and wood structure built with black bricks, with front and back sections, pavilions, gardens, and so on In the front, there are a stage in the center, surrounded by corridors and rooms, stone carvings and inscribed steles. In the back is the conference hall where the highest decisions were made by the financial industry of Ningbo in the old days. In front of the hall is a garden with a pavilion surrounded by trees and flowers. The round brick window is delicately carved with two coiling dragons. The Banking Guild Hall boasts a distinctive style integrating Chinese and Western styles. It is the only cultural relic building of the banking industry that is completely preserved in China.

---

[1]  A foundation set up by Yan Kangmao, a banker and entrepreneur in modern Ningbo.

# Chapter Thirteen

# The Drum Tower: Morning Bell and Evening Drum

Every morning and evening, the sound of bells and drums will ring over the city of Xi'an. People will see a team of "ancient warriors" marching through the Drum Tower Square, either in square or in line, attracting citizens and tourists to stop by for appreciation. The ancient city of Xi'an has restored the ancient ceremony of "morning bell and dusk drum", bringing a new connotation to the modern city.

According to *Cihai* (*Chinese Dictionary of Etymology*), "morning bell and dusk drum" was originally used to tell the time in temples. In the Northern Zhou dynasty (557), Yuwen Jue succeeded to the throne and made Chang'an (nowadays Xi'an) the capital. It was stipulated that the bell and drum be used to toll the hour for the royal family. Later, the practice carried on generation by generation and was spread all over the country, which common people relied on for time. After the Tang and Song dynasties, many scholars used the "morning bell and dusk drum" to refer to alarming words. For example, Li Xianyong of the Tang dynasty wrote in the poem "In the Mountains": "(In the mountains) the morning bell and dusk drum cannot be heard, and the bright moon and the clouds are lonely in the sky."

"On the watchtower, the drum and bugle are sounded to warn the army barracks." This is a verse by Chen Fu, who was a poet of the Yuan dynasty, which depicts the special function of the drum tower (also used as the watchtower) in Chinese history. In ancient times, the drum tower was equipped with the water clock (an ancient timer) and the nightwatchman's drums. Most of the time, the drum on the drum tower was beaten to announce the time; but in wartime, the drum tower served as a watchtower, shouldering the mission of protecting the city and resisting foreign aggression.

The most notable extant drum towers in Ningbo are Shunjiang Tower in Yuyao, the Drum Tower in Zhenhai, and the Drum Tower in Ningbo City.

Shunjiang Tower in Yuyao, also known as a drum tower, stands on the city wall. It was first built in the Huangqing reign of the Yuan dynasty (1312–1313) for the purpose of announcing the time. The building was destroyed in the third year of the Jiajing reign of the Ming dynasty (1522–1566), while rebuilt in the middle of the Wanli reign and equipped with the bell, the drum and *yunban* (cloud-shaped iron board, used for announcement of public notices). Later it was destroyed and reconstructed time and again. The existing building was rebuilt based on the original appearance in the eleventh year of the Guangxu reign (1885), with a double-eave gable-and-hip roof, seven bays wide and five bays deep. The stable and elegant Tongji Bridge and the simple and solemn Shunjiang Tower form an impressively imposing scene, which has become a symbol of the history and culture of Yuyao City.

Located in the east of the urban area of Zhenhai, Zhenhai Drum Tower was part of the military buildings in the ancient Zhenhai County. It was mainly used for lookout and time announcing. It was a relatively high building in the city in the ancient time.

The county of Zhenhai (formerly known as Dinghai) was originally built in the third year of Kaiping reign of the Later Liang dynasty (909), with a perimeter of only 450 feet. In December of the third year of the Jianyan reign of the Southern Song dynasty (1129), Emperor Gaozong Zhao Gou was chased by the Jin soldiers. He had stayed in Zhenhai for three days before leading his people to Changguo (i.e., present-day Zhoushan) and Wenzhou by sea. It is said that he had been paid obeisance to by the officials on the Drum Tower in Zhenhai.

In the twentieth year of the Hongwu reign of the Ming dynasty (1387), in order to prevent the invasion of *wokou* (Japanese pirates) and other pirates, Tang He, the Duke of Xinguo, expanded the Dinghai city to a perimeter of 1,288 feet and built Dinghai Wei ("wei" stands for the place where troops were garrisoned). In the twenty-ninth year of the Hongwu reign (1396), Liu Cheng, Chief Military Commissioner of Dinghai Wei, built the Drum Tower at the current site for watching the military situation. Covering an area of 500 square meters, the building is based on a foundation of bar-shaped stones, which is 5.9 meters high, 32.8 meters long and 16 meters wide. Stepping up the stone steps, there is a five-bay building equipped with facilities for timing and time announcing, such as a nightwatchman's drum, a bronze bell and a water clock. The drum was beaten to announce the time every day, allowing civil and military people to orient their lives around the time signals. On the building upstairs, Chinese traditional Four Seasons and 24 solar terms were posted. Downstairs is an arched passage shaped with 2-meter-long curved bar-shaped stones, and the top of the arch is five meters above the ground. At either end of the passage is a stone architrave with inscriptions, the south one inscribed with "朝宗古迹" (the historic site where people paid obeisance to the emperor), and the north one "东南屏翰" (the guard against the enemy in the southeast of China).

In the fifty-ninth year of the Qianlong reign of the Qing dynasty (1794), the county magistrate Wang Chengruo rebuilt the Drum Tower. At that time, the county had been renamed "Zhenhai", so the Drum Tower was also known as "Zhenhai Tower".

Ningbo Drum Tower is the south gate of the Zicheng City built in the first year of the Changqing reign of the Tang dynasty (821). It is the symbol of the establishment of city in Ningbo in the Tang dynasty.

At that time, Han Cha, Prefect of Mingzhou Prefecture, moved the government from Xiaoxi Town to the Three-river Junction. Centering around the area from Zhongshan Square to the Drum Tower, government offices were built, around which a timber fence was erected to form a city. Later, city walls were built with bricks and stones and Zicheng City came into being. At the south gate of Zicheng City is the present Drum Tower.

It is worth mentioning that when Wang Anshi, Grand Councilor of the Northern Song dynasty, a remarkable politician and a great reformer in the 11th century, ascended Drum Tower several times when he was the county magistrate of Yin County in the eighth year of the Qingli reign of the Song dynasty (1048). With people living a hard life on his mind and the sound of the drum and bell in his ears, he wrote an essay "Epigraph to the New Water Clock". In the essay, he resolved to manage political affairs "neither make haste, nor come late", just like the water clock. It seems that Wang Anshi was praising the water clock in the essay, but in fact, he was making a pledge to carry out a reform and eliminate the age-old malpractices. It was in Ningbo that he explored a set of reform ideas and accumulated some experience for the reform.

Emperor Gaozong of the Song dynasty, Zhao Gou renamed Drum Tower as "Fengguo Junlou Shrine" (the memorial temple in honor of the dedicated army) when he was in Mingzhou. It is said that he was chased by a large number of enemy soldiers and fled to the drum tower. When he was hiding there, he suddenly saw the five generals who had sacrificed their lives in the An-Shi Rebellion in the Tang dynasty, Zhang Xun, Xu Yuan, Nan Jiyun, Yao Chen and Lei Wanchun, who came to meet him dressed in armory and carrying a banner. Soon, Jin's army arrived, but they found nobody but cobwebs everywhere in the building and left. Zhao Gou, who thus managed to escape, later issued an imperial edict to confer upon the drum tower a new name "Fengguo Junlou Shrine", in which the statues of the five generals were erected for worship.

In the ninth year of the Xuande reign of the Ming dynasty (1434),

Governor Huang Yongding rebuilt the Drum Tower on the original site of the Tang and Song dynasties. He inscribed a plaque with " 四明伟观 " (Grand View of Siming) and another with " 声 闻 于 天 " (the sound of the drum and bell can be spread to the heaven) on the north, and had them placed at the center of the south and north of the Drum Tower upstairs. In the thirteenth year of the Wanli reign (1585) when the Drum Tower was about to collapse, the governor Cai Feiyi commissioned the reconstruction and changed " 四 明 伟 观 " into " 海 曙 楼 " (Haishu Building), which was derived from the verse by a Tang dynasty poet Du Shenyan: "Only those who left home to take official positions are sentimental to the changes in nature. The sun is rising from the sea, coloring the clouds in the sky. Willows are turning freshly green and plum flowers are blossoming red when I arrived in the south in spring by the river." Haishu Building, namely the building of sea sunlight, implies that "the waves are peaceful and the sea is safe" and that "from the sea the sun is rising".

In the Qing dynasty, the Drum Tower was reconstructed several times. The existing pavilion of the Drum Tower were built under the supervision of General Surveillance Circuit Duan Guangqing in the fifth year of the Xianfeng reign of the Qing dynasty (1855). In 1935, as proposed by the local people, a square Western-style lookout tower and an alarm post, made from cement and steal and measuring more than 6 meters in height, were built in the middle of the third story of the timber structure building, together with a bronze bell and a large modern mechanical standard clock on four sides, which were used either to tell the time or to sound the fire alarm. It was really a great idea to integrate the traditional architecture of the Tang dynasty with the Western architecture in such a unique way. In 2011, Ningbo Drum Tower was announced as one of the sixth batch of the Historical and Cultural Sites Protected at the Provincial Level of Zhejiang.

Ningbo Drum Tower had been ringing the bell and drum every day since the Tang dynasty, the harmonious sound traveling through the ancient city.

On every Chinese New Year's Eve since 1996, the bell has started to ring again on the Drum Tower to announce the end of an old year and the beginning of a new year. In1998, Ningbo municipal government started the renovation of the cultural scenic area of the Moon Lake, which brought hope for Drum Tower to restore the long, deep and exciting sound of the bell and drum. Since it is closely located to the south of Drum Tower, the Moon Lake looks graceful and beautiful viewed from Drum Tower, and the Drum Tower looks magnificent and glorious viewed from the Moon Lake. They are comparable to an old couple of time, watching the people of Ningbo City with affection.

The cultural scenic area of the Moon Lake was formed more than 1,000 years ago, which still retains a large number of ancient architectural structures of the Ming and Qing dynasties. In addition to Tianyi Pavilion, Lin's Residence and many other sites of cultural relics, there are also winding alleys around Baokui Lane. If you can hear the echoing sound of bells and drums in the morning and at night, you might think of the lines in Du Fu's poem "Visiting Fengxian Temple in Longmen": "On hearing the ringing of the morning bell just before being fully awake, I was startled and came to think over what life was all about."

# Chapter Fourteen

# Confucius Temples and Old Schools: Ancient Places of Schooling

Confucian culture has become the main body of Chinese traditional culture in the long history of China, and it still has a great influence even today. Confucius, the founder of Confucianism, grew from a scholar to being posthumously entitled as "Duke of Wenxuan, the Supreme Sage of Great Accomplishment". The Confucius Temple, a temple for people to worship him, has become a holy place in every Chinese county, and a building that must be set up in cities above the county level in ancient China. Confucius was deified and became a symbol of Chinese traditional culture.

Corresponding to Wu Temple (a temple in memory of a military saint, Guan Yu), Confucius Temple is commonly known as Wen Temple (a temple in honor of a philosopher, Confucius). Confucius Temple plays an important role in the history of ancient Chinese architecture. The one in Qufu is one of the three existing major ancient architectural complexes in China. It is also the origin of Confucius temples all over the country. The numerous Confucius temples in other places follow the layout and form of Confucius Temple in Qufu, reaching a number of 1,560 in total by the end of the Qing dynasty. In the fourth year of the Zhenguan reign of the Tang dynasty (630), Emperor Taizong ordered Confucius Temples to be built in schools around the country, resulting in a special type of ancient architectural complex—temple school architecture. The temple school architecture is a combination of the Confucius Temple and *xuegong* (school), i.e., the ancient Chinese local official school. Confucius Temple is the core building of *xuegong* complex, while *xuegong* is the basis of the existence of Confucius Temple.

Local temple schools are generally located inside the local provinces or prefectures, and counties. Their construction scale and standards are at very high levels in the local area. Local annals all over the country have records about their local temple schools, with school maps attached. Therefore, the local temple school is of vital importance in the history of Chinese architecture and education. For example, the Confucius Temple in Ningbo and the one in Yin County were specially documented in *The Annals of Ningbo* and *The General Annals of Yin County*, with detailed architectural drawings.

Among the Confucius Temples, the Imperial Academy Confucius Temple and Confucius Temple in Qufu are at the highest level, and the temple school of the prefecture is at a higher level than that of the county. In spite of the hierarchy, all Confucius Temples embody the common spirits and thus follow a set of common principles and rules in architectural composition and sacrificial activities.

From the architectural perspective, most Confucius Temples have Lingxing Gate, Panchi Pond, Dacheng Gate, Dacheng Hall, Dongxiwu (the side halls on the east and west of Dacheng Hall), Zunjing Pavilion, Minglun Hall, Jingyi Pavilion, Chongsheng Temple, Xiangxian Temple and Minghuan Temple. The architectural layout is mostly symmetrical on a clear central axis.

From the perspective of sacrifice, in addition to centering upon worshiping Confucius, the Confucius Temple also honors "Four Correlates", "Twelve Philosophers", and other scholars, local officials, and so on. The sacrificial rites of the Confucius Temple have also changed through times, but after the Tang and Song dynasties, a set of sacrificial rites dedicated to the Confucius Temple has gradually formed, namely "shidian" (a memorial ceremony by offering alcohol and food sacrifice).

"Shidian" is also known as "Ding Ji", which is scheduled to be held on the first *ding* day of the second month of each quarter (the 2nd, 5th, 8th, 11th lunar month) in a year.

According to statistics, most local temple schools were located inside the counties or prefectures where the government offices were located, and most of them were situated in the southeast or southwest of them. In addition to the prefecture-level temple school, often many county-level temple schools were also set in the city where the prefecture government sat. In the past time, there used to be eight Confucius Temple Schools in Ningbo, including those of Cixi, Yuyao, Zhenhai, Fenghua, Ninghai, Xiangshan, among others. In the city of Ningbo, there were Ningbo *fuxue* and Yin County School. Up to now, only the Confucius Temples in Zhenhai and Cixi (Cicheng) have survived. The two in Ningbo City and the one in Xiangshan no longer exist, while those in Ninghai, Fenghua and Yuyao still have some relics preserved.

The original Confucius Temple in Ninghai was built in the Kaiyuan reign of the Tang dynasty (713–741), located in the eastern suburb, now Dongguan Mountain. In the fourth year of the Qingli reign of the Song dynasty (1044), the first school was built and an official position—*jiaoshouguan* (professor in charge of education) set in the county. In the fourth year of the Jiayou reign (1059), the Confucius Temple and the school were integrated. After several relocations, in the sixth year of the Shaoxing reign of the Southern Song dynasty (1136), the county magistrate Qian Jun moved the Confucius Temple to the southwest corner of the county (now near Ninghai Hotel and Chengnan Primary School). The Confucius Temple was demolished in the 1970s, and now only Panchi Pond and Yuanqiao Bridge remained there.

After Fenghua was established as a county in the twenty-sixth year of the Kaiyuan reign of the Tang dynasty (738), the Confucius Temple was set up according to the rule originally at the eastern foot of Jinping Mountain. At the beginning, it was only a place for worshiping Confucius and offering sacrifices. In the Northern Song dynasty, a county school was set up in it, and it was moved to the city during the Shaoxing reign of the Southern Song dynasty (1131–1162).

The present Confucius Temple survived with Kongsheng (*lit.* Sage Confucius) Temple, which is the main building, Panchi Pond, and Kua'ao Bridge left. Kongsheng Temple was rebuilt in the eighth year of the Xianfeng reign of the Qing dynasty (1858) with several old cypresses planted in front. As the only temple-style architecture in Fenghua urban area, it has a gable-and-hip roof and double eaves, with upturned roof corners and *chiwei* decoration. Panqiao Bridge is a single bridge with a single stone arch spanning 3.6 meters and it has a total length of 17 meters. It is built with longitudinally laid block stones. On the wall of the bridge are four projecting carvings of lotus leaves. The carvings and the masonry method are relatively rare among the extant stone arch bridges in Ningbo.

Located in the east of Jiangnan Xuenong Experimental Primary School, Confucius Temple School of Yuyao was built by the county magistrate Xie Jingchu in the seventh year of the Qingli reign of the Song dynasty (1047), and its historical mission was not completed until the abolition of the imperial examination in the thirty-first year of the Guangxu reign of the Qing dynasty (1905). For more than 800 years, the temple school had fostered many talents for Yuyao, a "Famed Literature Land". Now the main hall and other buildings no longer exist except Panqiao Bridge, which was moved in 1985 to the back of the Youth Palace at the north foot of Longquan Mountain kept as a dry bridge without water flowing below. It is a single-arch stone bridge with a net width of 2.68 meters. Its name " 棂星桥 " (Lingxing Bridge) is engraved on the outside of the bridge railings, and " 巽水源流 " (where wind and water start and flow) on the other side.

In the twenty-sixth year of the Kaiyuan reign of the Tang dynasty (738), Mingzhou was established, and subsequently *zhouxue* (prefecture school) was also built in Xiaoxi (now Yinjiang Bridge) where the government was situated. In the fourth year of the Zhenyuan reign (788), Wang Mu, the Prefect, built

Dacheng Hall in the school. Therefore, the Confucius Temple of Mingzhou (Ningbo) also experienced the process of building the school first and then the temple, before the integration of the school and the temple.

In the first year of the Changqing reign of the Tang dynasty (821), the prefecture government moved from Xiaoxi to the Three-river Junction, and then Zicheng City was built, which laid the foundation for the development of Ningbo City for more than 1,000 years. The prefecture school was also moved there.

Since the second year of the Tianxi reign of the Northern Song dynasty (1018), *fuxue* had been located in the site of present-day Sun Yat-sen Square. In the eighth year of the Qingli reign (1048), Yin County magistrate Wang Anshi established the county school on the basis of a temple, which was later destroyed in the invasion of Jin's army. In the thirteenth year of the Jiading reign of the Southern Song dynasty (1220), the county school was rebuilt at the former site of the Xiweiguo Command Camp in Baoyun Temple (now the west section of Ningbo No.1 Hospital). It remained there through the Yuan, Ming and Qing dynasties and had been expanded. In the period of the Republic of China, the county school was abandoned. In the 1960s, the buildings of the county school were still well preserved, but now the main buildings have gone, leaving only the gatehouse. The gatehouse is a three-bay arch-style building in the shape of Chinese character " 八 " and with a gable-and-hip roof and four upturned roof corners. The *ruyi*-pattern bracket sets under the eaves are exquisite and completely preserved. Besides, there remains a Xianxue Street (County School Street) named after the county school.

In 1997, in order to construct the Sun Yat-sen Square, archaeologists carried out a rescue excavation of the Confucius Temple Site on North Jiefang Road, covering an area of 1,000 square meters. The main relics include the Site of Dacheng Hall, Panchi Pond, and so on. Today, there are remnants of a pier of Pan Bridge in the Sun Yat-sen Square, which was excavated from under underground and preserved in that excavation. It was covered with a glass roof and it had a sign set up inscribed with "Site of Panchi Pond Ningbo Confucius

Temple". It has now become a historical site for people to learn more about Ningbo *fuxue*. As early as in the period of the Republic of China, Zunjing Pavilion, a building with triple eaves and a gable-and-hip roof in the Confucius Temple School of Ningbo, was moved to Tianyi Pavilion Museum.

Zhenhai Confucius Temple is located in Zhenhai High School, and there remain Dacheng Hall, Dacheng Gate, Panchi Pond and Yanchi Pool. Dacheng Hall was destroyed in the twenty-fifth year of the Wanli reign of the Ming dynasty (1597), and was rebuilt by the county magistrate Ding Hongyang. In the twenty-second year of the Guangxu reign of the Qing dynasty (1896), Sheng Bingwei and others raised funds for a major overhaul. In 1937, the local people donated money and made a thorough renovation by taking apart its roofs and beams. It was transformed into a reinforced concrete structure, covering an area of 580 square meters. The one-story building faces south, with a width of five bays and a depth of four columns and nine purlins, surrounded by corridors. It is a structure of five-purlin beams and two-step beams in the front and at the back, with double eaves, a gable-and-hip roof, red rafters and black barrel tiles.

The Confucius Temple located in the center of Cicheng Town is now the only well-preserved Confucius Temple in Eastern Zhejiang. Many Confucius Temples that used to be even more spectacular than this one have been completely destroyed. It is only from some written documents and pictures that people can track down some clues of their previous existence. It is really rare that the Confucius Temple in Cicheng has survived the war and lasted for a hundred years.

As recorded in *The Annals of Cixi County*, in the first year of the Yongxi reign of the Northern Song dynasty (984), in order to cultivate talents, Li Zhaowen, the county magistrate, set up the county school 40 steps away from the county government. In the eighth year of the Qingli reign (1048), the county magistrate Lin Zhao moved the school to the current site. Wang Anshi, the Yin County magistrate wrote for a stele inscription "A Note on Cixi County School". Later he invited a knowledgeable local scholar Du Chun to be the school teacher. Through the wars and changes in the history, the Confucius Temple has been abolished and rebuilt repeatedly, and gradually formed its current scale.

Cicheng Confucius Temple adopts the layout of temple section in the front and school section at the back. The overall plan is arranged according to the central, east and west axes, neat and magnificent, reflecting the Confucian aesthetic standard of "beauty is neutralization and harmony".

The plan layout of the Cicheng Confucius Temple is compact but equipped with all functions. Compared with larger Confucius temples, such as Beijing Confucius Temple and Deyang Confucius Temple in Sichuan, both covering an area of more than 20,000 square meters, Cicheng Confucius Temple is only about one third of their size. But the tourists of Cicheng Confucius Temple feel as if it were larger than the other two, because it adopted some techniques like a change of scenery in each step of the journey, "borrowed scenery" (incorporating background landscape into the composition of a garden) and "multiple entrances", all of which serve to enlarge visitors' view sight. The designer of Cicheng Confucius Temple took advantage of the environment and made a successful design of the plan layout, which is similar to the layout of classical gardens of Suzhou.

Cicheng ancient architectural complex (Confucius Temple) was announced by the State Council as one of the sixth batch of the Major Historical and Cultural Sites Protected at the National Level in June 2006, and received the UNESCO Asia-Pacific Heritage Awards for Culture Heritage Conservation in 2009. It is the only architectural complex in Zhejiang Province that has earned this honor.

# Chapter Fifteen

# The Carving Art of Lin's Residence: A Wonder of Eastern Zhejiang

As a vital port on the southeast coast of China, Ningbo is located between mountains and seas, at the estuary of Hangzhou Bay and Yongjiang River, with abundant natural resources and convenient transportation. Thanks to generous gifts from nature, the diligent Ningbo people for generations have become good at trade, long voyage and thriving in other countries. There is even a saying that "no Ningbo businessmen, no business". In addition to its economic prosperity, Ningbo also attached importance to education and thus enjoyed prosperity in culture; numerous great scholars were born in Ningbo who ranked among the best in the imperial examinations and became important people in history. Ningbo people have been well aware of the relationship between businessmen, Confucians and officials. Under certain conditions, they skillfully realize the mutual promotion and transformation between the commodity economy and the *jinshen* (government officials) economy, thereby bringing a lot of wealth to the city. In order to honor and glorify their ancestors and their clan, the *jinshen* and wealthy businessmen spent a lot of money to construct large-scale buildings in their hometown, such as magnificent temples, solemn ancestral halls, grand residence, and elegant gardens, with sumptuous decorations and exquisite carvings. It is worth mentioning that the "Three Carvings" technology of wood carving, stone carving and brick carving is widely used for the screen wall, the gatehouse, the screen and other places, which are the best parts for the owners to show their status and for the craftsmen to display their skills.

Although the "Three Carvings" are just accessories for the ancient architecture, instead of the main body, their embellishment and decoration make the originally dull and empty buildings magnificent, gorgeous and full of life.

In particular, the carved items of the Ming and Qing dynasties are informative, vivid, true to life, which demonstrates the skillful craftsmanship of ancient Ningbo people in carving art.

Such ancient architectural structures fully decorated with the art of "Three Carvings" could be seen everywhere in Ningbo in the past. After so many years, few of them, the once-popular ancient temples and residences, decorated with lanterns and receiving banquet guests, have survived, and the vast majority of them have found themselves decaying and even on the verge of collapse. Among the extant buildings, Lin's Residence is well-preserved and deserves to be the representative of carving art in Eastern Zhejiang.

Lin's Residence is a large-scale, completely preserved and exquisitely carved traditional residential building in Ningbo urban area. It is typical of the residential houses in the middle and late Qing dynasty in Eastern Zhejiang in that the architectural layout displays a pleasant density, and the decorations are plain outside but beautiful inside. As the most concentrated, exquisite and rich place of brick, stone and wood carvings in our city, Lin's Residence is of great value to the study of carving art and architectural decoration art in Ningbo and even Eastern Zhejiang.

Lin's Residence in Zijin Lane at the south section of Zhenming Road, Ningbo, was built by Lin Zhongqiao and his brother Lin Zhonghua of the "Nanhu Lin family" in Ningbo, during the Tongzhi reign of the Qing dynasty (1862–1874). It is said that all the timbers used in Lin's Residence were purchased from Fujian Province. The house and garden were carefully and delicately designed by the Lin brothers. After several years of construction, the mansion with three gates, four sections, five *tianjing* courtyards, five-bay houses and two alleys was finally completed.

Lin's Residence covers an area of about 3,400 square meters. The main gate, the *yimen* gate, the screen wall, and the southwest courtyard are all arranged in the front section. The main buildings on the central axis face south, including the screen wall, the sedan chair hall, the main house, and the back building. The east and west wing rooms have double eaves. In front of the main gate is Zijin River. Over the river used to be Zijin Bridge. Behind the bridge is Zhenming Mountain.

The first gate is located on the east side of the building. Facing south is a towering brick arch-style gatehouse with an overhanging gable roof. It is about 7 meters high and 3.5 meters wide. The front side is inscribed with " 庆 云崇霭 " (auspicious clouds) up in the middle. There are two brick carvings of figures are on the left and right, with character stories in high relief below, and dragon patterns on the stone architrave. The column is engraved with a Chinese character " 寿 " (longevity), the patterns of *ruyi*, lotus and *wanfu* (the pattern of all happiness). The back side of the gatehouse is inscribed with " 春 风 及 第 " (very happy to pass the imperial examinations) and carved with the pattern of magpies bringing good news. Besides, approximately 26 different contents of carvings can be seen there. On the wall at the entrance there are varieties of carvings of diverse themes, showing superb carving skills, including "Eight Horses", "Grand Preceptor and Junior Preceptor", and "Phoenix Facing the Sun".

The *yimen* gate is a timber structure and faces east. People inside the house generally did not go out of *yimen* gate when they were welcoming the guest, especially the women. The visiting guests were allowed to enter *yimen* gate only with the permission of the owner, who would welcome the visitors there.

*Menzhen s*tones (the stone placed on both sides of the gate to hold and fix the gate pivot and support the gate frame) of the *yimen* gate are carved with patterns of Chinese character "回" and lotus flowers. Wooden columns and tie-beams are carved with vivid and unusual patterns such as grass dragon, lotus, *ruyi*, bat as well as fish jumping out of the spring water. The stone pier at the gate is a small stone lion in a lovely posture. On either side of *queti* there are two pairs of stone lions with *ruyi* pattern in openwork carving. The upturned eaves, *ang* and bracket sets are exquisitely made. The bracket sets are composed of one regular block and two minor blocks (used to support the bracket arm above) decorated with cloud patterns or water patterns. It is no exaggeration to say that where there is wood there is carving.

At the entrance of the *yimen* gate, we can see a tall and gorgeous screen wall. The upper section of it is carved with pictures of "Eight Immortals", "gods of happiness, wealth and longevity", "Saint Harmony and Saint Union" (symbolizing a happy marriage), "picture of nine old people" and "24 filial sons" by using the techniques of high relief, deep carving, and openwork carving. These carvings are surrounded by some other decorative patterns, such as exotic and rare flowers and herbs including ganoderma, orchids and daffodils. The screen wall of about half a square meter integrates the pattern of pavilions, mountains, rivers, boats, bridges, flowers, grass and clouds, displaying an elaborate multi-layer picture.

On the opposite side of *yimen* gate is a small garden named "Lanting" in the southwest of the residence, which comprises Lanting Pavilion, rockeries, ponds, and pavilion gardens in imitation of the text in "Preface to Lanting Pavilion Collection" by Wang Xizhi, a great calligrapher of the Jin dynasty, "here are lofty mountains, dense forests and slender bamboos, and a rushing stream flows around the pavilion". On the wall is inscribed with this text by Dong Qichang, a well-known calligrapher and painter of the Ming dynasty, followed by a postscript by Chen Jiru, a literatus and calligrapher.

The sedan chair hall on the central axis is a three-bay building with a single-eaved gable roof. The wing rooms on the left and right are used as the bedrooms for the sedan bearers or waiters, or as a small tea hall like

"Mingxuan". The internal and external beam frames, the wooden corbels and sill boards of the middle gateway are carved with diverse elements, including patterns of dragons, curly grass and clouds. On the walls under the eaves of the secondary rooms and side rooms on the east and west sides, there are extremely exquisite bar-shaped brick carvings such as the pictures of "Herding", "Farming and Weaving", "Harvest", and so on. The architrave of the gateway, the cornices, the windows, and the stepped gables are decorated with carvings on such themes as "Lady Burying Flowers", "Squirrel Stealing Peaches", "Good Luck and Good Fortune". The patterns on the brick windows between the secondary rooms and side rooms are quite regular, unified, and graceful.

The decorative patterns of wooden components in the front hall, the main room and the rear room are similar to those carvings on the roofs, including the images of phoenixes, curly grass and clouds. Unfortunately, the "Three Stars" or the "Three Gods" (Gods of Happiness, Fortune, and Longevity) on the roof ridge have been destroyed. In the middle of the *tianjing* courtyard between the front hall and the main room, there are two screen walls facing each other, on which under the eaves are embedded Chinese traditional cultural images such as "composing poetry while drinking in the boat on the autumn river".

The exterior look of the entire Lin's Residence is relatively simple, but the interior decorations are extremely complicated and sumptuous, showing the implicit and restrained Confucian character of Ningbo people. The main gate is arranged to the east side. Inside the mansion, there are many escape lanes, screen walls and high walls, reflecting the design of "hiding without revealing" and "winding paths leading to a secluded place" in the old mansions in Eastern Zhejiang. The building complex blends the art of wood carving, brick carving and stone carving very well, with diverse themes, exquisite layouts and simple but skillful knife techniques. There are nearly 250 carved patterns, including more than 180 stone and brick carvings, more than 50 wood carvings, and various types of decorative patterns on eaves-end tiles, gutters, and color paintings, which are distributed all over Lin's Residence. Therefore, it is reputated as the art museum for Three Carvings in folk houses in Eastern

Zhejiang.

There are many different kinds of carvings in Lin's Residence. In terms of the carving materials, there are wood carvings, stone carvings and brick carvings. Based on the type of patterns, the carvings can be divided into six categories—animals, plants, geometric shaped, historical stories, texts and utensils. The common carving patterns in Lin's Residence are an expression of auspicious folk art, with folk customs and traditions as its core. The form and implication of the decorative patterns are reflected in and also enhanced by each other. We might as well get to learn more about them.

The motif of "Everything Goes Well". This good wish is expressed by means of the homonyms of "*shi*" ( 事 , a thing)—"*shi*" ( 柿 , persimmon) and "*shi*" ( 狮 , lion). In Chinese culture, the plant persimmon and the animal lion both convey auspicious meanings. For example, the lion symbolizes prosperity and fertility. Therefore, in Lin's Residence there are small stone lions used for gate piers and four pairs of *ruyi* lions in openwork carving.

The motifs of "Fortune and Longevity" and "Five bats Bringing Longevity". Generally, "fortune" or "blessing" is represented by the animal bats ("bat" is the homonym of "blessing" in Chinese) or the plant fingered citrons (alias Buddha's hand, as the fruit looks like the Buddha's hand; "佛" is the homonym of "福"). "Longevity" is mostly symbolized by the patterns of longevity peaches, longevity rocks, the character "寿", pine trees, cranes and so on. The patterns of bats and "寿" can be seen everywhere in Lin's Residence.

The motif of "Three Friends in Winter", namely pine, bamboo and plum blossom. The pine tree is an evergreen tree, so it is taken as a symbol of perseverance, strong will and immortality. The bamboo symbolizes elegant, free and unrestrained demeanor. Despite its capability to survive the bad weather, it does not take too great pride in itself, but displays the qualities of modesty and integrity which are also embodied in a Chinese gentleman. The plum blossom is loved by people for its icy beauty, its bravely blossoming in snow, and its fragrance and elegance, which symbolizes a noble character. The pattern of "Three Friends in Winter" on the carved brick gatehouse of Lin's

Residence demonstrates the owner's attitude towards friends and friendship.

The motif of "Picture of Eight Horses". The eight horses in question are originally the mounts of King Mu of the Zhou dynasty. "Their feet do not touch the ground; they have wings; they run faster than birds and travel thousands of miles at night", so they are also known as "eight heavenly horses", which can fly up to the sky and down into the sea. "Picture of Eight Horses" in Lin's Residence is composed of eight horses in high relief and rolling clouds. Each horse has a different look and posture, either standing or sitting, making a long hissing or making a soaring leap. The carving skills are perfectly displayed, and it implies the meaning of "The horses are running fast in the spring, pleased and proud".

The motif of "Friendship Between Gentlemen", also known as friendship between orchids and ganoderma. It is a metaphor for the social interaction with friends of good character. There is a saying that "being in the room with orchids and ganoderma for a while, you do not smell its fragrance because you have become a part of it". There are usually orchids, ganoderma and reefs in the pattern, since "reef" and social interactions are homonyms in Chinese.

The motif of Magnolia in the Breeze. Magnolia in Chinese culture symbolizes purity and elegance, and the combination of magnolia and peony is called *yutang fugui*, which means wealthy and prosperity.

The motif of *guqin* (a plucked seven-string Chinese musical instrument), Go, calligraphy and Chinese painting. Having always been loved by literati and painters, they used to be the elements commonly adopted in carvings.

The motif of Eight Immortals. The most common is the figure patterns. Since the Ming and Qing dynasties, the Eight Immortals' weapons are used to allude to the Eight Immortals.

The motif of national beauty and natural fragrance, which refers to Chinese peony, known as the "Queen of Flowers" in China. It is a famous flower with a unique color and fragrance, usually symbolizing wealth and glory.

The pictures of Herding, Farming and Weaving, Entertaining, and

Harvesting in Lin's Residence are carved on a bar-shaped brick with a length of about 4 meters and a width of 0.3 meters. The four pictures are separated by pictures of flowers and birds. There are more than 60 figures in these pictures, including men and women, old and young.

The picture of "A Lady Burying Flowers", 0.35 meters long and 0.25 meters wide, is located at the end of the stepped gable. The well-dressed plump lady in the picture is carrying a hoe and a flower basket on her shoulder, and had a deerlet on her right.

Patterns of tailed dragon and grass. Since the Ming and Qing dynasties, the carving patterns of dragons and phoenixes in varied forms, flowers and grass have been commonly used as cultural symbols, and they are also the most common in Lin's Residence.

The serial carvings of People Composing Poems While Drinking in the Boat on an Autumn River. Passing through the dark lanes and arriving the place between the front hall and the main room, you will find yourself in the fourth courtyard. The most amazing carvings in Lin's Residence are the brick carvings on the two opposite screen walls in the *tianjing* courtyard here. This type of brick carved screen wall is also rare among architectural structures in Jiangsu and Zhejiang. The screen wall is about 2.8 meters high, 7.5 meters long and 0.32 meters thick, on which is a 7.5-meter-long and 0.25-meter-wide bar-shaped brick carving. The bar-shaped brick carving is composed of eight images in the style of a picture-story book, narrating story of people composing poems while drinking in the boat on an autumn river. The images are separated by carvings of animal and plant patterns. Among them, one image depicts smiling young scholars and beauties on boats. The river waves, mountains, two boats, the items in the boats, pavilions in the distance, and the boatmen are all delicately and vividly depicted in full details. The beauty is pouring water into the river on the boat, and the young man is standing in the boat behind, staring at the beauty. In another image, a beauty is peeking out from the door, half of her body hidden behind, the lines of her dress carefully and delicately carved. The tiles on the roof are also shown in great details. The carvings are just like a series fine realist paintings.

Motifs including "Four Blessings Coming Together" and "Incomparable Fortune and Longevity" are used for the four square hollowed-out brick windows below the bar-shaped carvings. The windows are basically the same in size and shape.

From this pair of brick-carved screen walls, we can see that the residence was built in such a unique and original way that the building could not only function well as a dwelling place but also present the beauty of culture. The texture and color of the bricks create a sense of authenticity and familiarity. The carving art of Lin's Residence is worthy of being entitled as a unique wonder in Eastern Zhejiang.

# Chapter Sixteen

# Ancestral Halls in Ningbo: The Bond of the Family Clan

## I. Reasons Why Ancestral Halls Were Needed

Walking across the urban and rural areas of Ningbo, you can often see one or two unusual houses with towering walls and upturned roof corners. Inside the exquisite houses, you can see carved and painted beams and columns. Despite some differences in scale, these buildings have one thing in common, that is, there is a plaque right above the gate with large golden characters "...'s Ancestral Hall" on black background.

In the feudal society of ancient China, the concept of family is highly valued. Often a family or several families of the same surname lived in the same village, and many of them built their own clan temples to worship their ancestors. These clan temples are generally called ancestral halls, which include the clan's ancestral hall, branch ancestral hall and family's ancestral hall. The name "ancestral hall" first appeared in the Han dynasty when ancestral halls were built in the graveyard, called graveyard ancestral temples.

Zhu Xi of the Southern Song dynasty set up the rules of ancestral halls in his *Family Rituals*. Since then, the clan temple was named ancestral halls. At that time, the construction of ancestral halls was differentiated by hierarchy, and civilians were not allowed to build ancestral halls. During the Jiajing reign of the Ming dynasty (1567–1572), "civilians were permitted to set up a clan temple together with other clans of the same surname". Later it was stipulated that only the surnames any family member of which had been granted an award or conferred dukedom on by the emperor can be called "clan temple", while the rest were called the clan's ancestral hall.

As a distinctive traditional cultural form in China, the existence of ancestral halls is inseparable from the deep-rooted family concept for thousands of years. In the old days, an ancestral hall was usually built in a village inhabited by a large clan, and its scale was often decided by the wealth of the family. If a clan was rich and powerful, its ancestral hall was mostly exquisite, which could become a symbol of the family's glory.

Ancestral halls are the places where people offer sacrifices to ancestors or distinguished people of the past. They served for many purposes. In addition to the purpose of "worshiping the ancestors", they were also used for other purposes such as marriage, funeral, birthday, and other big events. Moreover, they were also used as places for discussions of important affairs of the clan, and also as places for gatherings of the clan members.

The culture of ancestral halls has a history of thousands of years in China. From the initial purpose of offering sacrifice to ancestors, it gradually evolved into a bond to maintain the relationship of the same clan. By researching the various plaques, genealogies and other materials preserved in an ancestral hall, we can have a better knowledge of the emergence, development and migration of the clan in the long history, and we can also get a glimpse of some social and cultural information of various historical periods. Nowadays, in urban and rural areas of Ningbo, ancestral halls in many places have been renovated and well protected, becoming places for future generations to pay tribute to their ancestors and recall the clan's history.

# II. Representative Examples of Ancestral Halls

Ningbo has a long history, and a large number of clan's ancestral hall buildings have been preserved in different counties and districts of Ningbo. Among them, Qin's Branch Ancestral Hall and Zhang's Ancestral Hall in Haishu District, Xie's Ancestral Hall of the First Ancestor in Simen, Yuyao, and Cai's Ancestral Hall in Yinzhou District can be the representatives.

## 1. Qin's Branch Ancestral Hall, The Pinnacle of Exquisite Craftsmanship

Qin's Branch Ancestral Hall was incorporated into Tianyi Pavilion Museum as one of the fifth batch of Major Historical and Cultural Sites Protected at the National Level in June 2001. The temple was built by Qin Jihan, a descendant of the Qin family and a former banking tycoon in Ningbo, and was completed in 1925.

Here is a legend about the origin of Qin's Branch Ancestral Hall.

Around 1921, there was a young man from Ningbo, Qin Jihan, who was engaged in business in Shanghai. He met a German businessman who dealt in pigments. As we know, Germany was the most developed country in the chemical industry in modern times. At that time, the vast majority of pigments in China were imported from Germany. Right after the end of World War I, however, the German businessman was eager to return home to visit his families, so he discussed with Mr. Qin about the possibility of exchanging his pigments on hand for some money to return home. In order to solve his friend's urgent need, Mr. Qin obtained his shop of pigments at a low price. Soon, the war in Europe broke out. So the import of pigments was suspended for a time, while the textile industry in Shanghai was booming. Due to the short supply, the price of pigments suddenly soared by dozens of times, and therefore Qin made a huge fortune and rose to wealth overnight. After that, the Qin family successively invested in banks and real estate industries with low risks but rich profits, and their wealth snowballed.

In honor of his ancestors, on returning to his hometown in Ningbo in 1923, Qin Jihan proposed to rebuild Qin's Ancestral Hall in Zhangqi Lane, Zhenming Road when he found it in very poor condition. In the eye of the elders in his clan, it was preposterous for a young man to propose the construction of an ancestral hall just because he had some money, and they felt embarrassed and humiliated. Besides, Mr. Qin was not a direct descendant of the Qin family. Finally, the two sides reached a compromise that he was not allowed to build the ancestral hall, but could build the branch ancestral hall. Having no choice, Qin Jihan made up his mind to build a branch ancestral hall far outshining an ancestral hall. After investigating, he chose Mayancao beside the Moon Lake, a good place according to *fengshui*, and a beautiful, quite place distanced from the hustle and bustle of the city. He selected Hu Rongji Construction Factory for the important task of building a Qin's branch ancestral hall. Therefore, the best carpenters, masons, lacquerers, stonemasons and other master construction craftsmen in Ningbo at that time were gathered and then the construction waited to commence at an auspicious time. It is said that in order to build this ancestral hall, the participant workers of Hu Rongji Construction Factory were so dedicated to the construction that they worked day and night, even skipping meals and sleep sometimes. They worked hard to build a "brand project" of their own. If there was a slight problem in quality during the construction, it would be redone. For example, a stone stool was replaced due to slightly poor quality, even though it has no obvious defect and might be mistaken for a sample. After two years of efforts, by 1925, an unprecedented ancestral hall has finally been erected by the Moon Lake.

Qin's Branch Ancestral Hall faces south, with the east garden of Tianyi Pavilion Museum on the north, and Mayancao River on the south. The building complex has three sections in total, with a rectangular plan. On the central axis are the entrance hall of the screen wall, the stage, the main hall and the back building, with wing rooms are on the east and west sides. All these form a large-scale timber structure building complex, covering a total area of nearly

2,000 square meters. The construction took a total of more than 200,000 silver coins. The decoration of the building integrates the characteristics and styles of traditional local crafts of previous dynasties. What is impressive about it is the use of a large number of red-lacquered gilded wood carvings. In addition, boxwood carvings, stone carvings and brick carvings were also employed. Such scale and quality are extremely rare in ancestral halls among the common people. As a representative work of architecture in the early period of Republic of China, it is a valuable example for the study of architecture and carving arts of the early 20th century.

## 2. Xie's Ancestral Hall of the First Ancestor, the Top One among Those of the Ming and Qing Dynasties

Xie's Ancestral Hall of the First Ancestor is located in Houtanghe Village, Simen Town, Yuyao City. In the Ming dynasty, Xie's clan in Simen Town was divided into 18 branches, each of which had its ancestral hall. Nowadays, these branch ancestral halls have disappeared, and only "Xie's Ancestral Hall of the First Ancestor in Simen" has still been well preserved. It is located in Houtanghe Village, locally known as the "Grand Ancestral Hall". It was first built during the Zhengde reign of the Ming dynasty (1506–1521), and the construction was proposed by Xie Qian and undertaken by Xie Pi. Xie Pi, the second son of Xie Qian, ranked the third in the highest imperial examination, or *tanhua* in the eighteenth year of the Hongzhi reign (1505), and was finally appointed as *zuoshilang*, or the Left Vice Minister of Personnel and *Zhangyuanshi* (the official in charge of the academy affairs in Hanlin Academy) of Hanlin Academician. On the architrave of the main gate are the characters " 四门谢氏始祖祠堂 " (Xie's Ancestral Hall of the First Ancestor in Simen) inscribed by Xie Zheng in the Ming dynasty. Xie Zheng was the eldest son of Xie Qian, who was granted the title of *Zhongshu Sheren* due to his father. Later, he worked in Wenyuan Pavilion, editing the history of emperors, before being promoted to Vice Director of the Ministry of Rites and granted the title of Chief Deputy Envoy to Liao. Xie Zheng was also a master of four styles of calligraphy, namely the regular script, cursive script, clerical script and seal script.

The ancestral hall comprises three sections of buildings and fifteen bays in total. In the central bay of the first section is the memorial tablet of the two remotest ancestors of the clan who first moved to Yuyao. In the east and west side bays are the memorial tablets of the next eighteen generations. In the central bay of the second section are the memorial tablets of the three *Taifu* (Grand Mentor in charge of civil affairs), namely, Xie An of the Jin dynasty, Xie Shenfu of the Song dynasty, and Xie Qian of the Ming dynasty. In the east and west side bays are the tablets of the Revered Xian, (i.e., Xie Xuan) and the Revered Ruhu (i.e., Xie Pi), respectively. The first and the second sections of buildings are single-story palace-style houses, spacious and magnificent. The third section is a multi-story building, the central bay of which houses the tablet of the Revered Bodeng, the ancestor who first moved to Jiangnan in the Jin dynasty, and the east and west side bays are for the worship of the Revered Shengwu and the Revered Daoyuan. The third section has an architectural style quite different from the first two sections. The wood carving is dealt with extremely meticulous care, with a distinctive style of Huizhou architecture. There is a stone stele embedded on the wall downstairs on the east side inscribed with round and full lines of calligraphy, but no writer had signed it. Research found that it was written by Xie Jiashan, a *juren* in the Qing dynasty, when he was ten years old. There is a single-story house with eight bays on the east side of the ancestral hall, which used to be the residence of the person who attended the ancestral hall before the founding of People's Republic of China. The house in the east is the shed for the storage of lanterns. Simen Lantern Festival was once the grandest Lantern Festival held in Yuyao. It originated in the Yuan dynasty and has a history of more than 600 years. The one-story building with five bays in the west of the ancestral hall is a barn. The entire ancestral hall covers an area of 2,200 square meters. Its large scale and complete preservation are rare in Eastern Zhejiang.

The ancestral hall was burned down by the Taiping Army (army of Taiping

Tianguo, or the Taiping Heavenly Kingdom) in the first year of the Tongzhi reign (1862), and rebuilt by the patriarch Xie Yingsong in the third year of the Guangxu reign (1877). Only the stone structure of the main gate of the outer wall and the architrave inscribed with " 四门谢氏始祖祠堂 " are the original objects at the initial construction in the Zhengde reign of the Ming dynasty (1506–1521).

### 3. Zhang's Ancestral Hall in the Ming Dynasty, the Representative of Early Ancestral Halls in Ningbo

Zhang's Ancestral Hall, located at No. 70 of Qingshi Street, Haishu District, is one of the few existing buildings of the Ming dynasty in Ningbo, and is also a representative of the early ancestral hall buildings in Ningbo. It is now among the Historical and Cultural Sites Protected at the Municipal Level of Ningbo. According to the survey, in the Ming dynasty, celebrities such as Zhang Shipei, Zhang Xikun and Zhang Xihuang lived on Qingshi Street, and the ancestral hall should have been built by their clan people. Being a place where Huang Zongxi used to give lectures in his later years, it was one of the main bases for dissemination of the Eastern Zhejiang School.

The ancestral hall faces south, and on the central axis are an entrance hall and a main hall. It covers a total area of 465 square meters and a construction area of 335.5 square meters. The entrance hall has five bays, three columns and five purlins, and the central bay is the entrance gate. The gable adopts the style of *guanyindou*. The main hall has a width of three bays and with two lanes. The central bay uses five rows of columns and nine purlins. The side bays with the lanes have seven rows of columns and nine purlins. The main hall is a mixed structure of the column-beam-and-strut framework and the column-and-tie-beam framework. The beam frame is made of huge wood materials, decorated in a simple yet elegant fashion. The drum-shaped column base reflects the architectural style of the Ming dynasty.

## 4. Cai's Ancestral Hall, A Rare One with Male and Female Ancestral Halls Combined

Located in Panhuoqiao Village, Yinzhou District, Cai's Ancestral Hall was founded in 1588. The existing buildings were built in 1870 during the Tongzhi reign of the Qing dynasty. The architectural layout, construction techniques and practices represent the characteristics of Ningbo traditional ancestral hall buildings. The building complex covers a large area. In particular, the main hall of the back section is nearly nine meters high, and the materials used are in huge size, which was extremely rare in Ningbo and even Zhejiang. In addition, Cai's Ancestral Hall is rare example in China where the male and female ancestral halls were combined, which is the first in Zhejiang. It proves that since 1870, the Cai's clan has broken the feudal rule that women were not allowed to enter the ancestral hall. This practice was very rare at that time.

# III. Features of Ancestral Halls in Ningbo Land

As the buildings with special purposes, ancestral halls are strictly regulated in location, layout and composition. Besides, influenced by regional culture and clan culture, ancestral halls in different regions, or of different clans, have their own characteristics. The buildings of ancestral hall in Ningbo share the following characteristics.

## 1. Location Selection Principle of "Harmony Between Humanity and Nature"

People associate *fengshui* of ancestral halls with a clan's prosperity, so the site selection of a new ancestral hall should be very careful. Generally, it is necessary to pay attention to the dragon vein (representing the path of *qi* flow) and the source of *qi* (a type of force or energy that is believed to affect us physically and mentally). It is also important for the building to be located with back to mountain and facing the water, to have a spacious and square courtyard, and not to expose the inlet and outlet of water. According to *fengshui*, "there must be *qi* in a place surrounded on both the left and the right". Therefore, orientation is of great importance in selecting the location. Generally, a building should face south or east. Sometimes people decide on different orientations according to specific conditions of dragon vein. In short, the location, orientation, form and layout of ancestral hall buildings must take into account factors linked with family prosperity and development, and the principle of "harmony between humanity and nature", that is, to choose a place with very auspicious *fengshui* features.

Some ancestral halls adopted the plan layout of Panchi Pond of the Confucius Temples. Through adopting the ceremonial facilities in the past temple school people hope that more descendants of their clan would be successful candidates in the imperial examinations. Besides, people also believed that the pond symbolizes accumulation of wealth and facilitates fire fighting. For example, Qin's Branch Ancestral Hall was chosen to be built at Mayancao beside the Moon Lake, a beautiful and quiet place with very auspicious *fengshui* features. Mayancao is the largest pond in front of the ancestral halls in Ningbo. Wu's Ancestral Hall in Xidian Town, among others, also have Panchi Pond and Panqiao Bridge in the front courtyard on the central axis.

There are also images of "four treasures of the study" in the pool in front of Xie's Ancestral Hall of the First Ancestor in Simen. From north to south, they are "ink, inkstone, mountain-shaped pen holder and white paper", which were all artificially made. "Ink" is represented by a small island shaped like a round ink-cake. The "inkstone" is a deep channel, which never dries up all year round. "The mountain-shaped pen holder" was made of three earth piles. "White paper" is symbolized by The river is a symbol of white paper. And the "brush" is represented by trees on the hill.

## 2. The Scale and Decoration of the Ancestral Hall Determined by Financial Power

Since the organization and layout of ancestral hall buildings are regulated, the overall layout is basically the same. An ancestral hall is generally composed of the square in front of the gate, the stage, the gate, the enclosure, the *tianjing* courtyard, ancestral room, the worship hall, the dormitory room, and the auxiliary rooms. The size depends on the financial condition of the clan.

For example, Qin's Branch Ancestral Hall has three sections of buildings, with a rectangular plan. The north-south central axis consists of the screen wall, the gate hall, the stage, the main hall and the back building, with wing rooms on the east and west sides, forming a large-scale timber structure complex with a total area of nearly 2,000 square meters.

The ancient architectural decorations in China are mainly color paintings and carvings, both of which have a long history and distinctive national characteristics. Color painting is not only for decoration, but also for protecting the wood, and carvings add vividness to the building. Architectural carving technology began with clay sculptures in the middle and late primitive times. During the Sui, Tang, Song and Yuan dynasties, it witnessed a profoundly epoch-making development. In the Ming and Qing dynasties, a complete set of traditional craftsmanship were formed. Brick carving, wood carving and stone carving had their own characteristics. Stone carving and brick carving were mainly used for exterior decoration, mainly in the foundation, the gate, the halls, the top of the gable, the ridge, and so on, and wood carving was mainly used for interior eave decoration.

The stage caisson ceiling of Chen's Ancestral Hall in Yingjiashan Village, Taoyuan Sub-district, Ninghai County is exquisite with colored paintings, and some patterns use shading and perspective methods, which give them the illusion of three dimensions.

Zhang's Ancestral Hall is a building of the Ming dynasty with brick carved *taimen* with colored paintings, which might well be the earliest colored paintings in ancestral hall buildings in Ningbo.

Plaques, column couplets and steles are the highlights of the ancestral hall architecture. Some inscriptions on them are concise but meaningful, some interesting but full of remarkable literary talent. Plaques and columns written by celebrities are particularly precious, such as the plaque of "Yanyi Hall" hung in the middle of Ancestral Hall of the First Ancestor in Sunjiajing, Cixi during the Ming dynasty. The plaque of Xie's Ancestral Hall in Simen is still the original one built in the Zhengde reign of the early Ming dynasty (1506–1521). The stele with the inscription "唐始祖太傅公赡田碑记" (stele inscription in the farmland to commemorate the first ancestor in the Tang dynasty, the Revered Taifu) was erected in the sixtieth year of the Kangxi reign of the Qing dynasty (1721) in Yunlong Town, Yinzhou District, which tells the glorious history of the ancestors of the Huang family.

**3. The Symbolic Building of the Ancestral Hall Indicating the Owner's Identity**

There was one rule in the ancient regulation of building ancestral halls: Only those having an official position above the fourth grade in the clan were allowed to erect tall flagpoles in front of the ancestral hall. After the reform of the Republic of China, flagpoles were allowed to be erected for centenarians and those who had made significant contributions to their hometown after doing business abroad and becoming rich. Therefore, flagpoles can be found in front of the gate of ancestral halls of wealthy and celebrated clans.

The memorial arch right in front of the ancestral hall is a symbol of noble character and respectable status. The more exquisite the carvings on it, the more it exhibits the ancestor's noble character and the later generations' respect. There are stone arches and wooden arches, and in Ningbo the majority are stone arches. On Datong Road in front of Xie's Ancestral Hall of the First Ancestor in Simen, there were originally two memorial arches, the east one inscribed with "太傅流芳"(the Grand Mentors will be glorious through the history), and the west one with "东山并秀"(these prominent figures from Xie's clan, referred to as Dongshan Hall, are equally distinguished). The architraves were inscribed by Lü Ben, one of the three Grand Secretaries of Yuyao in the Ming dynasty. Unfortunately, both arches have been destroyed.

The larger ancestral halls have stages. The stages in Southern and Northern China are different in style. The stages in the south are more delicate and elegant, while those in the north are more formal and stately. Obviously, the stage has become the most important part of the ancestral hall. The large-scaled ancestral halls in Ningbo are basically equipped with stages. For example, the stage of Qin's Branch Ancestral Hall is still the most exquisite stage in modern Ningbo.

# Chapter Seventeen

# The Earliest Western Architecture in Ningbo

In 1840, with the outbreak of the Opium War and China's reduction into a semi-colony, the traditional Chinese architecture system received impacts from all fronts and was forced to modernize. The earliest foreign architectural style that entered China was the "colonial-style" architecture, which marks the beginning of the modernization of Chinese architecture.

This architectural style originated from the imitation of the local Indian native architecture by British colonists. In the 17th century, when European countries expanded outward on a large scale, British colonists came to Asia. The colonists designed a veranda-style building by combining the styles of Indian architecture with veranda and British architecture which suits a temperate zone, so as to adapt to the tropical climate in India. This type of architecture is distinctively characterized by the indispensable veranda, which serves as the most comfortable space in the residence. Some have it on one side only in the front of the building, while others have it on all sides. On the veranda, colonists would have their light meals and tea, smoke, chat, do their knitting, read books, play chess, or enjoy a nap.

This kind of veranda-style architecture was popularized with the expansion of colonialism. In China, despite the climatic differences between China and India, especially in the lighting this architectural form, as a symbol of colonialism, still enjoyed popularity for more than half a century from the Opium War to the early 20th century. Initially, it was adopted for office buildings such as the consulates. With the evolution and maturity of this type of building in form and function, it was gradually adopted for various buildings, such as commercial firms, guild halls, schools, foreign firms, banks and residential houses, commonly used in the second half of the 19th century. People also created varieties of designs to enrich the forms of veranda-style architecture, which played an important role in promoting the modernization of Chinese architecture.

On January 1, 1844, Ningbo Port was officially opened. Britain, France, the United States, Germany, Russia and other Western countries came to trade with China. In 1850, Britain and other countries forcibly delimited a large area of land along the North Riverbank, as the "residential area and commercial port area for foreigners", where those foreign countries began to  set up their consulates.

The winter in Ningbo is quite cold, but this did not hinder the rooting and development of this veranda-style architecture in Ningbo. From the opening of the port to the beginning of the 20th century, Ningbo had built a large number of veranda-style buildings. The main types that have survived include the following. First, religious, cultural and educational buildings, such as the former sites of the Catholic Abbey, the veranda-style lobby of the Bishop's Office, the office building of Zhedong Middle School and the building of Christian church on Huaishu Road. Second, public and municipal buildings, such as the former sites of Zhejiang Customs, the British Consulate and Ningbo Post Office. Third, buildings of foreign banks and firms, such as the former sites of the office buildings of the Swire Group, British Merchant Firm and Hefeng Yarn Factory as well as the former railway station master's office. Fourth, former residences of celebrities, such as Xie Hengchang's Private

Residence, Fu Qianyue's Room of Fu's Residence, a Ningbo businessman Zhou Jinbiao's Residence, and Weng Wenhao's Former Residence.

Judging from the existing buildings, the veranda-style buildings in Ningbo include four types: single-side columned veranda, double-side columned veranda, triple-side columned veranda and four-side columned veranda. The plan is mostly square, rectangular or in the shape of capital letter "L", with relatively simple functions. The rooms mostly adopt the interspersed style, integrating functions for work and for residence. The main rooms are equipped with fireplaces, which have good natural lighting and ventilation. The veranda can be single-story or double-story. Most were double-story, while those early-built ones are single-story. The triple-story and multi-story verandas had not been adopted until the introduction of steel-reinforced concrete in the early 20th century. The columns of a "beam-column" veranda are composed of one or two types, the latter adding variety and liveliness to the design. Most of the columns are built with bricks, and their plans can be rectangular, square and circular. The columns are usually covered with facings of bricks with plastering, or with pointing but without plastering.

From the form of the facade, the veranda-style buildings in Ningbo fall into two major categories, namely "arcade style" and "beam-column style". The veranda of the arcade style comprises a row of outer columns connected by continuous arches, which adds variety and magnificence to the veranda. The veranda of a beam-column style takes the columns of the veranda itself as the main decorative component, and the beams between the columns are connected by coping, which makes it more dignified and solemn.

The walls of the veranda-style buildings are made of Chinese traditional black bricks, which are much thinner than the Western red bricks. The material used to stick bricks together is not mortar or cement, but clay. The cracks between the bricks is covered by mortar, as if the bricks had been held together by it. As the overhanging cornice of the eaves requires certain strength of the cementing material, bricks there are cemented with mortar. The cornice's brackets are made of bricks, and then plastered with thick mortar to create the curving outline.

It should be noted that the function of veranda-style buildings has changed since it began to be used for residence. The consulate generally used to serve as both office and residence, but the residential building had only the residential function. From the plan of Weng Wenhao's Former Residence, it is obvious that a living unit is surrounded by verandas on three sides. This kind of veranda-style villa is one of the main forms of villa architecture in modern China. In addition, Xie Hengchang's Private Residence displays more local traditional features in decoration, construction, techniques and designs, forming a modern veranda-style architecture integrated with the local decoration characteristics in Ningbo.

Among the existing modern buildings in Ningbo, the most typical "colonial-style" buildings are the former sites of Zhejiang Customs and the British Consulate.

In October, 1843, the British authorities sent the Consul Robert Thom and an interpreter to Ningbo, and set up the "British Consulate in Ningbo", commonly known as the "British Mansion", temporarily located in a folk house at No. 1, Yangjia Lane, Huaishu Road, Jiangbei District. After Ningbo was officially opened, British Mansion was moved to No. 56, Baisha Road in 1880, which is the present-day former site of British Consulate.

It is said that the Consul Robert Thom had made two contributions in Ningbo. First, due to his preference of playing mahjong, it was he who first spread Chinese Mahjong overseas from the Old Bund of Ningbo. The other is that he selected and translated a part of the sixth chapter of *A Dream of Red Mansions*, and included it in an English book *The Chinese Speaker* published in Ningbo. This is the earliest English translation of this novel. Robert Thom may be regarded as an early transmitter of Chinese culture in Ningbo.

In June 1934, the British Consulate in Ningbo was closed down, and the British Consul in Shanghai, John Fitzgerald Brenan, then transferred the house to the government of Yin County to serve as a relief home. After the founding of new China, the original Consular's residence and staff houses of the

British Consulate were demolished, leaving only the main building (the office building). After years of war and changes of the times, it remains as the only consulate building in Ningbo.

The building of British Consulate faces east. Several large square columns made of black bricks support the front porches of the two stories. In front of the porch is a bottle-shaped guardrail, which looks plain yet elegant. The hip roof is covered with Western tiles, on which are four square fireplace chimneys.

The buildings in the former British Consulate were arranged surrounding a courtyard with lush trees. The existing office building of the British Consulate is a two-story building, symmetrical along the east-west central axis, with verandas on all sides. The verandas on east, north and south sides are wide, while the one on the west side is comparatively narrow, and was separated as a storage room. The main entrance is located in the east, facing the inner courtyard. On the first floor, the main entrance, lobby and main stairs are arranged on the central axis. The secondary stairs are hidden in the northwest corner. The inner corridors connect rooms on the same floor. The main rooms are arranged on an east-west axis, and the bathroom is located in the west. The layout of the second floor is similar to that of the first floor.

The facade of the building adopts the classical combination of colonnades and veranda, a result of the eclecticism in architecture prevailing in the West in the 19th century. The design is suitable for the humid and hot climate here as it functions well in sheltering from the rain and ventilation as well as preventing the heat. Besides, it has a beautiful appearance with quite excellent artistic effects.

The climate is also a factor to determine the size of windows, as windows directly affect the ventilation of buildings. Therefore, with the large windows, the facade of the consulate building demonstrates good ventilation and lighting. The compound window composed of two or three single windows have a complex design and many moldings, which looks very prominent on the facade.

The architectural decoration style is jointly influenced by Western architectural trends and Chinese traditional culture in the modern time, showing a dominance of Western styles and local Chinese characteristics. As the representative of each country, consulates must have their own characteristics and therefore demonstrate great cultural differences from each other. As a result, they were all carefully designed and constructed at great cost, with distinctive designs and forms that are highly ornamental and representative of the country.

Zhejiang Customs, founded in 1861, is located at No. 198 Zhongma Road, Jiangbei District, Ningbo. Its former site is about 20 meters away from Yongjiang River in the east, about four meters away from the Christian Church in Jiangbei in the south. About 200 meters further south, there remains the building of Senior Deputy Office of the former Zhejiang Customs, which is a Western-style two-story brick house and serves as both office and residence. It is located adjacent to Zhongma Road in the west, and about six meters away from the Ningbo Shipping Building. On the Site of Ningbo Shipping Building, there used to be the General Office Building and the Tax Department Mansion of Zhejiang Customs.

The existing former site of Zhejiang Customs is one of the buildings for office and management of the Tax Department of the former Zhejiang Customs. Facing east by south, the building is a brick and wood structure with three floors and a loft. It has verandas on two sides. It has a rectangular plan and covers an area of 1,067.80 square meters. Its entire width is 15.10 meters, the entire depth is 18.44 meters, and the height from the ground to the roof is 16.05 meters. The rooms of the first floor are arranged in L-shape, and three main rooms are located behind the eastern colonnade, which used to be the offices for examiners of the New Customs, the Port Affairs Section and the Inspection Section of Zhejiang Customs respectively. In the southeast is a wooden staircase, which is the main access to the second floor. In the southwest is a stockroom, and in the northwest are the other management rooms. The second and third floors used to be the residence of the Chief Procurator of Zhejiang Customs. The room division is basically the same as the first floor, but the west side is divided into several smaller rooms. The fourth floor is a loft with eight simple rooms. The exterior walls are brick walls without plastering, but were pointed with cement.

The pitched roof is a hard hill roof covered with square tiles, with the front slope longer than the rear one. The front of the roof is equipped with two skylights, and there are two high Western fireplace chimneys in the front and rear respectively. The east side of the building is a colonnade with six columns. Half of the south side that connects with the east side is a colonnade with three columns while the other half is a wall with windows. The square brick columns are treated with the technique of *mojiao* and topped with neat and powerful Corinthian capitals. The verandas on the second and the third floors have columns connected by bottle-shaped wooden railings that look graceful and delightful. The cross-section of the columns gradually shrinks from the first floor to the third floor, highlighting the effects of layering and rhythm of the building facade. Concrete stairs are set outside the back wall, which are connected with the back doors on the second and the third floors, and can be used as the emergency exit. The rails of the stairs are made of cast iron. On the west side of the back wall on the first floor is a side door. The outer wall has a string course comprising two layers of red bricks for each floor. The lintels of doors and windows are decorated with red brick arches. At the two-thirds of the columns are two layers of red bricks, and the column tops are also decorated with red bricks. All these indicate that at that time people have begun to pay more attention to creating aesthetic effects with color and material texture. Red bricks are used for a limited part on a large area of black brick walls to form a contrast in color, producing a strong visual and aesthetic effect.

The building emphasizes its practicality, but its interior decoration is exquisite, concise and lively. The joists of each story are covered with wooden floors, which are all three-centimeter-thick Oregon pine, slotted and spliced. The non-bearing walls and ceilings are built with wattles mixed with mud and coated with plaster. Basically, the places between the ceiling and walls are inlaid with multi-layer moldings, and the center of the ceiling is decorated with several circle patterns. The doors and windows are equipped with three-dimensional frames. Each wodden door has five rails, ten grids and a pin tumbler lock. Inside the veranda are a row of four double glass doors, each comprising five rails and ten grids. Outside the veranda are detachable wooden louvre doors. Inside the rectangular double windows are four-grid glass doors; outside are the window shades to block the sun and wind and to embellish the house. Each room or hall is furnished with a fireplace and closets. The interior staircase is wooden spiral one equipped with intricately designed bottle-shaped railings.

# Chapter Eighteen

# Modern *Shikumen* Architecture in Ningbo: "Living Fossil" of Ningbo Traditional Folk Customs

"The royal *kumen* is resided by imperial members; the stone *kumen* houses common people; no Ningbo native old enough have never walked into the building." This folk song was popular along the North Riverbank in Ningbo before the founding of People's Republic of China. The stone *kumen* mentioned is *shikumen* building, which was one of the major residential buildings of Ningbo people in modern times.

The origin of *shikumen* buildings can be traced back to the Zhou dynasty. As recorded by *Records on Examination of Craftsmanship*, "among the five gates built specially for the Son of Heaven, the first one is *gaomen*, and the second is *kumen*". In an explanation of "Shu" (explanations to the notes), "It is said that the *kumen* in the State of Lu has a similar form to the *gaomen* of the Son of Heaven." It was regulated by the palace building standards that the emperor's palace should be designed with five gates, dukes' or princes' residences with only three gates. Therefore, the first gate of dukes and princes should be *kumen*, whose construction level could only be similar to the second gate of the emperor's five gates. Although the State of Lu was a vassal state, the shape of its palace outer gate, or *kumen,* had a form equivalent to that of the Son of Heaven, namely *gaomen*, indicating the power of the State of Lu and the decline of Zhou dynasty at that time.

After the mid-19th century, early *shikumen* houses built with traditional Chinese column-and-tie-beam timber structures and using brick walls for load-bearing emerged in the Old Bund of Ningbo. Though the owners of these houses had been exposed to Western architectural culture at an earlier time, the deep-rooted Chinese traditional culture still played a dominant role, so the main building body still showed strong spatial characteristics of traditional folk houses in Jiangnan. In terms of layout, by adopting the design of Western townhouses, they display the combined characteristics of Chinese and Western architecture. Since there are many gates in this kind of folk houses and the outer gate frame is made of stone, people then called it "*shikumen*". This type of building in Ningbo is characterized by the local Meiyuan's red rock of Yin County that Ningbo people generally preferred to use for gate frames, and the black-painted thick wooden gate leaves and a pair of copper gate rings.

During the years of Taiping Rebellion, the rebellious army of Taiping Tianguo and the Qing army had a half year long tug of war in Ningbo, and a large number of people moved to the foreign residential area in North Riverbank to escape from the war. The house price soard, as there were too few houses yet too many people. Seeing huge profits, many local and foreign banks began to set foot in real estate business. At the end of the 19th century, a large number of new houses of *shikumen* style were built along the Old Bund in Ningbo.

The emergence of *shikumen* architecture is not entirely a historical accident, but seems to be a necessity of urbanization. The traditional courtyard-style residential buildings were gone forever, and the living space were compressed, but people were rewarded a new modern life style, as epitomized by wide roads, gas, street lights, running water, telephones, as well as new sets of values, tantalizing business opportunities and charms in the foreign residential area. Ningbo people at that time were amazed by these achievements of modern civilization. For example, with regard to the new gas lamp, also known as *shuiyue* lamp (*lit.* water moon lamp), at that time, as was reported by the newspaper *German Businessmen in Ningbo* on December 28, 1898, "Its flame is as bright as the moon, with far more blazing light than the candles".

The early 20th century was the heyday of *shikumen* architecture. In the alleys extending in all directions, small stores, warehouses, taverns, restaurants, banks, newspapers, bookstores, printing houses, and manufacturing companies all based themselves at *shikumen* buildings, which made a small society with all kinds of industries.

After the founding of the Republic of China, the extended families fell apart, because they were shackled by complicated ethics and thus were no longer suitable for urban life full of competition and pressure. As a result, the middle and late *shikumen* buildings fit for single immigrants and independent families came into being. Emerging from the three-story houses with three bays on the first floor, the *shikumen* buildings with two stories and two bays on the first floor or with one story and one bay expanded. had its overall scale greatly expanded. Also, the brick and wood structure with reinforced concrete is employed. Though with a smaller space, the new *shikumen* buildings have more varieties of designs and forms.

After the 1930s, with the arrival of the worldwide economic depression, the heyday of *shikumen* architecture had passed. In order to reduce the burden of rent or make money by selling the house, many *shikumen* residents rented out their spare rooms. Some low-level *shikumen* landlords even divided the house into smaller rooms, or built attics over them. *Shikumen* was no longer the exclusive paradise of the middle class, but gradually became the most popular residential house. Their residents included employees and managers of Chinese and Western enterprises, owners and compradors of small—and medium-sized businesses of various industries, as well as a variety of craftsmen and freelancers. They were the mainstay of citizens, who were a mobile but still stable group. The spaciousness, warmth and poetic atmosphere that *shikumen* *was* once endowed with vanished.

Today, when people talk about *shikumen* architecture, what comes to their mind first must be those in Shanghai. However, by reviewing the history of *shikumen* architecture and comparing the early *shikumen* residence in Shanghai and the modern architecture in Ningbo in terms of plan layout, door and window decorations, structure and building materials, it is safe to conclude that the modern folk residence of *shikumen* style in Ningbo is one of the main sources of the *shikumen* architecture in Shanghai, as it emerged and became the major form of early modern architecture in Shanghai when Ningbo businessmen immigrated to Shanghai.

As can be seen on the modern architectural block on the North Riverbank, modern folk houses in Ningbo are characterized by an integrated Chinese and Western styles, with a great many Western elements used. Their influence can be discovered in the decorative techniques of early *shikumen* buildings in Shanghai. Data show that red bricks have been used in a large number of *shikumen* buildings in Shanghai, while traditional black bricks have mainly been used in the modern buildings on the North Riverbank. It suggests that the modern folk houses in Ningbo appeared at an earlier time than in Shanghai.

There is no doubt, though, that the late modern architecture in Shanghai was more influenced by the architectural culture of imperialist colonial countries.

Whenever we walk into a *shikumen* building, it is like entering an architecturally decorative hall in various forms and styles. Entering the gate, we can see the sitting room in the middle, which is generally equipped with detachable floor-to-ceiling windows facing the *tianjing* courtyard. There are wing rooms on the east and west sides, which generally served as a study or for miscellaneous use. In the middle of the second floor is also a lobby, with wing rooms on both sides, both serving as bedrooms. Behind the main house and the rear *tianjing* courtyard is the attached house, which is generally used as the kitchen or storage room. The rear *tianjing* courtyard has a well. The whole house has a gate in the front and at the back as well.

*Shikumen* buildings owned by wealthy families mostly use such decorations as wood carvings, brick carvings, stone carvings. Among them, wood carvings are considered to be the most typical of folk houses in Jiangnan, but most of the wood carvings in *shikumen* buildings tend to be simplified. Wooden moldings and wooden railings with low relief could be mass-produced and became the most popular decorative components at that time. Red bricks also appeared, and were used together with black bricks for the same wall of the same building, producing a very special decorative effect. The combination of black and red bricks is a major feature of modern architecture in Ningbo. Meanwhile, cement was also used as decorative components, such as for the door relief, railings and so on.

The gate of *shikumen* is composed of a door frame, one or two door leaves and a lintel. In the early stage, the door frame was mostly made of stone without complicated lintel decoration. In the later stage (after 1920), it could be made of stone, brick and cement. The door frame and the lintel were highly decorated, some with multiple moldings, and some with pilasters imitating the Western classical style on both sides.

The lintel is the most fantastic part of *shikumen* buildings, also the most important part for decoration. According to Chapter "Records of Tang Dynasty" in *Comprehensive Mirror for Aid in Government*, when the noble consort Yang was in favor, a folk song got around that "Don't be delighted if you gave birth to a boy, and don't be sad if it is a girl, because the decoration of lintels depends on whether you have a daughter today". Hu Sanxing from Ninghai in the Southern Song dynasty noted: "Anyone who paid a visit at the house of Yang would be amazed at the wide span and magnificence of the lintel. It is said that the family of Yang rose to glory because of a daughter." It indicates the relationship between the lintel and social status, highlighting the importance of external decorations of the lintel. Influenced by the Western art of architectural decoration, *shikumen* gates in the later period often used geometric patterns such as triangles, semicircles, arcs, trapezoids or rectangles as the outer contour of the lintel, which generally had reliefs, Baroque patterns of the Renaissance period, traditional Chinese auspicious patterns or characters, or some mixed Chinese and Western patterns. The decorative patterns of these *shikumen* buildings reflect Shanghai-style mentality of the modern Ningbo people—open to new things and able to integrate things old and new, home and abroad.

As a typical product of the integration of Chinese and Western architectural cultures, Ningbo *shikumen* architecture, which emerged in the mid-19th century and flourished in the 1920s, accounted for more than half of the residential buildings at that time. To this day, there are still old Ningbo people living in *shikumen* buildings along the Old Bund. *Shikumen* architecture, a typical type of "modern folk houses in Ningbo" with both Chinese and Western architectural styles, has left its mark in the history of modern architecture in Ningbo.

The specific construction year and location of the first *shikumen* building can no longer be verified. The early *shikumen* buildings have vanished from history. Today, some well-preserved ones include Yu's Residence, Xu Ronggui's Residence, Jin's Residence, Zhong's Residence, Liu Sihai's Residence, Wang Chuihua's Residence, and Yan's Residence.

Yu's Residence is located at No. 1–No. 2, Shijun Lane, Jiangbei District. It is a modern *shikumen* building with a *sanheyuan* (a courtyard with buildings on three sides). The main building faces south. The outer wall is relatively high. It was made of the traditional brick walls without plastering, and there are two Western-style wooden windows on the gables. The gate on the wall is a modern *shikumen* gate with a combination of Chinese and Western styles, its lintel in the shape of semicircle, and the interior decorated with regular geometric patterns. The stone *queti* over the door frame is carved with patterns of plum blossom, orchid, bamboo and chrysanthemum in relief. The two-story, three-bay main building is equipped with two wing rooms and an outer corridor. The beam frame is of the column-and-tie-beam structure without any carvings. No. 2 of Shijun Lane was once attached to the house at No. 1 of Shijun Lane. Its gate also adopts the style of *shikumen*, whose lintel is decorated with three big characters " 迎春坊 " (the lane where spring is welcomed).

The owner of Yu's Residence was a modern businessman in Ningbo. Around the 1920s, he opened a store to sell goods from all over the country, located on Zhongma Road on the North Riverbank. It has the store in the front and a workshop at the back. After he gained fortune and fame, he built the *shikumen* building near the store so as to improve the living condition.

Xu's Residence is located at No. 17 of Xinma Road, with the gate facing the road. The two-story building adopts the *shikumen* design and terrazzo floors. The gate is in the form of semi-circular arch, and the facade is decorated with some regular geometric patterns. The exterior wall is composed of brick walls without plastering. The house has a gable roof with small black tiles and a column-and-tie-beam structure, and the ground paved with black stone slabs. Built with traditional timber structures, the main building has five bays and two levels.

Xu Ronggui was the earliest owner of Xu's Residence. According to *Comprehensive Dictionary of Businessmen of Ningbo Origin*, in 1890, a sailor named Wang Baocang opened a shop named Dexing Coppersmith to repair foreign locks, fire extinguishers, and so on After he passed away, his apprentice Xu Ronggui inherited his business. In 1900, Xu expanded his business scope, changed the name of the store into Shunji Machinery Factory, and launched machinery repair business. In 1924, Shunji Machinery Factory joined the "Machinery Association of the Great Republic of China". In 1951, the factory was purchased by the state and changed into State-owned Shunji Machinery Factory, which later became an important part of Ningbo Power Machinery Factory.

Jin's Residence is located at No. 1 to No.3, the 6th Lane, Shengbao Road. It consists of three parallel *shikumen* buildings, with similar architectural structures and a plan layout of a *sanheyuan* style. The *shikumen* gate made of terrazzo and bricks is decorated with patterns of semicircles, squares and other geometric figures, and the gate frame is also made of terrazzo. The main building has a width of five bays and two wing rooms. The front porch has bottle-shaped wooden railings. The front end of the porch floor is carved with *ruyi* pattern, curly grass pattern and other patterns. The building has a gable roof covered with small black tiles, and the gable top made of cement is an imitation of *guanyindou* style. On the gable there are modern Western wooden windows. The beam frame is of the column-and-tie-beam structure.

The original owner of this residence is unknown. It was sold to a businessman surnamed Jin before the People's Republic of China was founded, who used to own a department store in Japan.

Zhong's Residence was owned by a senior staff member of Hefeng Yarn Factory before liberation, who used to be a cashier. The residence, located at No. 3, Lane 10, Shengbao Road, is a building of *sanheyuan* (a courtyard house with buildings arranged on three sides around the couryard and one wall on the main gate side) style. The main building is three bays and two lanes wide, with a gable roof covered with small black tiles and a column-and-tie-beam structure. It has front porches. The second floor is made of wood, and the bottom floor is paved with black stone slabs. The front porch on the second floor has iron railings with flower patterns, which are beautifully made, with geometric patterns of arc, circle, fretwork, rectangle. The front end of the floor is decorated with patterns of curly grass, rhombus and fretwork. The *queti* of the porch bears the exquisite patterns of curly grass and *ruyi* in openwork carving. The stone column bases are carved with patterns of plum blossoms and *ruyi* cloud. There remain some colored paintings on the top of the gables and the enclosing walls, and at the foot of a gable lies a black rock inscribed with " 泰山石敢当 " (*lit.* rocks of Mount Taishan are able to take on heavy responsibilities) with *fengshui* implications.

Liusihai's Residence is located at No. 5–No. 7, Daici Lane. It is composed of two *shikumen* buildings, gardens, ponds and tea rooms. There is a path between the two *shikumen* buildings, and the entrance of the path also follows the *shikumen* design, which adopts a brick arch with a pointed arch with cement sculptures on it. The plan layout and structure of the two *shikumen* buildings are similar, both of the *sanheyuan* style. The main building adopts a column-and-tie-beam structure. The building has a front porch of two floors high. In front of the porch are wooden railings decorated with rectangular geometric patterns. The gable roof covered with small black tiles is made of cement while imitating the *guanyindou* style. There remains part of the colored paintings on *chitou* (the component at the upper part of the gable jutting out of the gable to support the eaves) and the enclosing walls. The slightly square pond is surrounded by iron railings. The tea hall is two stories high and four bays wide, built beside the pond. Its roof adopts a Western-style wooden frame and is covered with Western tiles. In the front of the second floor are a wooden railing and a full-length glass window with wooden lattice. On the first floor is a corridor with cement railings decorated with geometric patterns such as round and rectangular.

This building was owned by Liu Sihai, a modern businessman in Ningbo, who owned a wide range of businesses related to shipping, wharves, coal mines and so on, and had been to Japan for trade several times.

Wang Chuihua's Residence was once owned by a senior employee of Ningbo Zhengda Match Factory. It is located at No. 13, Lane 36, Xinma Road. It is a delicate small Western-style house. The main building is three bays wide and two stories high. Its facade has a special shape, with a flat central bay and the side rooms in the form of half-hexagon. There is a veranda on the central bay of the second floor, which is surrounded by a cement railing. The roof adopts a modern Western-style wooden frame covered with Western cement tiles, and the ground is made of terrazzo. The outer wall is decorated with rectangular patterns.

Yan's Residence house is located at No. 3–No. 14, Deji Lane, Jiangbei District. It consists of the main building, several rooms on the right, and a garden on the left. The main building faces south, with a plan layout in *sanheyuan* style. Facing Deji Lane, the gate is a brick gatehouse with a plaque which reads "Chang'an Yongkang" (*lit.* well-being forever). In the garden of plum, the stone *queti* is carved with patterns of plum blossom, orchid, bamboo and chrysanthemum. The double gate leaves are nailed with iron sheets and decorated on the inside with the pattern "four bats encircling longevity", which embodies good wish of blessings and longevity. The main building is two stories high, five bays wide and five bays deep, with a column-and-tie beam frame. The top of the veranda has a round ridge roof of three-level wave ceiling. It has hooks for lanterns still kept there on the beam. The *yueliang* beam is also carved with patterns of peony, curly grass, animals, plants, and human figures. The stone bases of outer eave columns adopt the design of *ruyi* and *gualeng*. The central and side rooms are all tiled. The lower plate of the windows is made of terrazzo, with a smooth surface. The interior plaster ceiling is decorated with the pattern of plum blossom in the center. The two wing rooms are also two stories high and one bay deep, with a column-and-tie-beam structure. The cement gable roof in an imitation of the *guanyindou* style is covered with small black tiles. The back building has a flat roof, which serves as a balcony that is accessible. The four-bay roof is encircled by railings.

Yan's Residence was owned by Yan Zijun, son of Yan Xinhou, a representative of the modern Ningbo Commercial Group. Yan Zijun, with the assumed name of Yibin, came from Cixi. After Yan Xinhou's death in 1906, Yan Zijun inherited his father's business. In addition to managing Yuanfengrun Draft Bank, he undertook Yuantong Customs Official Bank due to his close relationship with Cai Naihuang, Shanghai *daotai* (circuit intendant, supervisor of special administration zones directly subordinated to central government agencies). In 1908, he participated in the establishment of Siming Bank and Ningshao Shipping Company. With his business activities covering Shanghai, Beijing and Tianjin, among other cities, he has served as a board member of Shanghai General Chamber of Commerce for many times.

# Chapter Nineteen

# Modern Bridges: The Link Between the Past and the Future

After the industrial revolution in the mid-18th century, the production and casting of iron provided new construction materials for bridges. However, cast iron is not a good bridge construction material because of its poor impact resistance and weak tensile strength, which means it is easy to break. In the middle of the 19th century, with the development of Bessemer steel making process (converter steel making) and the open hearth steel making process, steel had become an important material for bridge construction. Its high tensile strength and good impact resistance, especially the emergence of steel plates and rectangular rolled section steel in the 1970s, has made it possible for manufacturers to assemble the bridge components in the factory. Conditions were prepared for the steel to be increasingly widely used and for the civil engineering to make its first leap. Later, with the invention of high-strength steel, the steel structure boomed. Therefore, the span range of bridge structures increased from a few meters or tens of meters of the masonry and wooden bridge, to hundreds of meters or even kilometers of the steel bridge, creating the miracle of building bridges across rivers and straits.

At the beginning of the 18th century, cement was invented by mixing and heating lime, clay and hematite. In the 1850s, steel reinforcement was placed in concrete to make up for the poor tensile strength of cement. In 1867, reinforced concrete was invented, leading to the second leap in civil engineering, which made it possible to build reinforced concrete bridges with long spans and diverse forms. The development and application of such artificial materials as steel, cement concrete and prestressed steel reinforced concrete (PSRC), as the symbol of modern bridges, has promoted the scientific and technological revolution of modern bridges.

From the thirty-third year of the Guangxu reign of the Qing dynasty (1907) to the first year of the Xuantong reign (1909), China built a steel bridge on the Yellow River in Lanzhou, Gansu Province, namely Lanzhou Yellow River Bridge, with the steel imported from Germany. Its construction basically marks the end of China's history of ancient bridge construction with wood and masonry as the main materials. Since then, China's bridge construction has marched into a new historical period.

With numerous rivers and dense water networks, Ningbo is a famous water town in Jiangnan and a "kingdom of bridges", creating a rich and colorful culture of ancient bridges. After the Opium War, with the introduction of foreign cultures, bridges in Ningbo experienced fundamental changes. In addition to the construction of traditional stone bridges, Ningbo has built many other bridges with the use of modern Western materials and construction technologies.

Based on the construction materials, the bridges in modern China are classified into wooden pontoon bridge, stone bridge, iron bridge, steel bridge, and reinforced concrete bridge. The existing modern bridges in Ningbo are mainly steel bridges and reinforced concrete bridges. Wooden pontoon bridges are only found in historical records but have not survived; stone bridges basically follow the traditional architectural style; no iron bridges have been found in historical records so far, nor have their examples been found.

# I. Lingqiao Bridge

Ningbo Lingqiao Bridge was built almost at the same time when the city of Ningbo was established. The history of Lingqiao Bridge can be traced back to the Tang dynasty. In the third year of the Changqing reign (823), Ying Biao, Prefect of Mingzhou prefecture, built the first wooden pontoon bridge in the history of Ningbo across Fenghua River near the Three-river Junction, in an effort to relieve the burden of business travelers. Sixteen boats were connected in a row with strips of bamboo, and then boards were laid on them. The bridge was 55 *zhang* (a Chinese measurement of length, about 3.3 meters) long and 1.4 *zhang* wide. Two years later, due to the wide river and strong current there outside the east ferry gate, it was moved to today's site. When it comes to the origin of the its name, according to historical records, when the bridge was being built, a rainbow appeared in the sky. Therefore, the bridge was named "Lingxian Bridge", also called "Lingjian Bridge", later "Lingqiao Bridge"[1]. It was renamed "Dongjin Pontoon Bridge" in the Song dynasty, or called "Laojiang Bridge" (*lit.* old river bridge) among the folk people, because another pontoon bridge was built at the end of the Yaojiang River near the Three-river Junction in the late Qing dynasty, called "Xinjiang Bridge" (*lit.* new river bridge), and the change was to distinguish the new from the old.

In the fifth year of the Qianning reign of the Tang dynasty (898), after Huang Sheng, Prefect of Mingzhou, built Luocheng City, "Lingqiao Gate" at the east end was named after the bridge. Li Yesi, a poet and litterateur in the early Qing dynasty, wrote in the "Yundong Zhuzhi Ci": "The boards of the Dongjin Bridge across the river float, / Holding sixteen boats in one row. / Thousands of people riding through, / The kindness is done by Ying Biao we shall know."

---

[1] Many took the appearance of the rainbow as a sign from heaven. The Chinese character "灵" in the bridge's name means "epiphany".

As a scenic spot at that time, Lingqiao Gate was also highly praised by literati. Shu Dan of the Song dynasty wrote in his poem entitled "On Lingqiao Gate": "The high building stands in the clear wind over the river; / The flag of the Ziluo River waves in the sunset. / In the summer rain the swirling brooks rage in the valley; / In the evening fog the land is getting hazy. / Through the layered walls we hear joyful voices, / From a distance we see the lovely and wild landscape. / Beautiful verses travel wide and far, / Which were written by Du Fu the genius." Wang Gen of the Song dynasty described it in "The Night View from Lingqiao Gate", "Gentle wave is in harmony with the mild drizzle, / enveloping thousands of houses in purple green. / The rainbow across the water is elegant and ethereal; / The Cowherd and the Weaving Girl were separated by the hazy Milky Way. / Gods are arriving at the Three Holy Mountains by way of Mount Chu; / Human beings singing songs are wealthy and want nothing. / Bearing the flag of victory, they can't wait to be back on their country road; / Enveloped by wind and moonlight, the boats are taking fishermen home."

During its existence of more than 1,000 years, the pontoon bridge has been broken and even destroyed many times due to natural and man-made disasters. The repeated disasters suffered by the pontoon bridge also brought inconvenience and sufferings to the people in Ningbo too. How they yearned for a solid and durable fixed bridge that is resistant to severe weather!

The idea of transforming the pontoon bridge into a fixed bridge started in the late Qing dynasty. By 1922, Chen Shutang, a local resident, had drawn up a plan for the reconstruction of the Laojiang Bridge and had it sent to all the authorities in Ningbo. This detailed reconstruction plan marked "the beginning of the reconstruction of the river bridge", as noted in *The General Annals of Yin County*. In the same year, a businessman Ying Minghe wrote letters to the prestigious old gentlemen in the city to suggest the reconstruction. After the meeting, Ningbonese Association in Shanghai sent a letter to Ningbo General Chamber of Commerce and Yin County Council. In September, Some German engineers were invited to Ningbo to conduct the survey and draw up the construction plan, with a project cost of 300,000 silver dollars. In early

November, the Yin County Council decided to establish the Engineering Bureau. After that, the proposal of bridge reconstruction was specified. The Ningbonese Association in Shanghai held a meeting in mid-December and established the "Preparatory Office for the Reconstruction of Ningbo Laojiang Bridge". But finally, it turned out fruitless due to lack of funds.

In late August 1926, a major disaster occurred on the pontoon bridge. After two consecutive days of heavy rain, the flood broke out in the mountains at the upstream of Fenghua River, accompanied by the spring tide of the East China Sea surging from the Three-river Junction, finally causing the the chain of the pontoon bridge to break and the bridge body to collapse. At that moment, people on the bridge rushed to escape, but more than 30 people fell into the water. Only three people were saved; the rest lost their lives. The tragedy led to heated discussions among Ningbo people not only in Ningbo but also in Shanghai regarding the reconstruction of the Laojiang Bridge. In the middle of September, Yue Zhenbao, Yan Kangmao, who were Ningbonese in Shanghai, and Luo De, a Westerner, came to Ningbo for survey. The first preparatory meeting was held in mid-October, and it was decided that Zhang Shenzhi, Yan Kangmao, Yu Guifang and other three people be in charge of the preparatory meeting. On November 5, a preparatory meeting was held at Ningbo General Chamber of Commerce, where 60 preparatory members were selected and the sponsors pledged to make the donation on the spot. Later, due to the Northern Expedition, the project was suspended again. Following the success of Northern Expedition, the situation is eased and the economy recovered, and therefore the pontoon bridge failed to meet the needs of industrial and commercial development. More and more people from all communities called for the reconstruction of the bridge. In 1931, Yue Zhenbao, Zhang Jiguang, Zhang Shenzhi, Zhu Quantong, Jin Tingsun and some other Ningbo people in Shanghai relaunched the "Preparatory Committee for the Reconstruction of Ningbo Laojiang Bridge" on August 1 of that year with the support of all parties.

After its establishment, it had offices set up both in Shanghai and in

Ningbo, with 20 members in Shanghai. The Director of the Preparatory Committee was Yue Zhenbao, and the Deputy Director was Chen Rongguan. It had 16 members in Ningbo, with Wang Wenhan as the Director, Yan Kangmao and Xu Yongsheng as Deputy Directors. Shi Qiuzang, the then Technical Director, namely engineer, of the construction project, said in his article *Memories of the Construction and Salvage of Ningbo Lingqiao Bridge*: "Given the old government's lack of public funds, it had to rely on donations. To raise donations, it was necessary to have foreigners take charge of the project in the old society, using foreign materials and operating under the name of foreign firms. Therefore, after discussion at the Shanghai Preparatory Meeting, A. F. Gimson, an Englishman from the Bureau of Works of Shanghai International Settlement, was invited as the consultant engineer for the bridge construction, with building materials bought from Siemens Ltd., a German firm, and the pile driving contracted out to A. Corrit Co., a Danish firm. In fact, the foreigners rarely came to Ningbo. The actual designer, the constructor and workers were all Chinese. I was entrusted by the county magistrate Chen Baolin as the Preparatory Committee member to be in charge of construction, supervision and audit. When the foreign experts came to Ningbo for inspection, I was the interpreter at the banquet. But it was a shame to be put at the mercy of the foreigners." In October 1933, the bridge construction project was put out to tender in Shanghai. Siemens China Co. won the contract with a bid price of 486,774 silver dollars. The steel beams were provided by M. A. N. Factory. The pile driving and concrete works were subcontracted to A. Corrit Co., Denmark, and the painting works were contracted to China Engineer Co. These parties have undoubtedly played an important role in guaranteeing the quality of Lingqiao Bridge, though the actual operation was all done by Ningbo people including those from Shanghai.

As for the bridge type, the Preparatory Meeting rejected the original proposal of a reinforced concrete bridge (which features two piers in the middle to make three arches, similar to the present Liberation Bridge), and adopted the design of "steel through arch highway bridge with three hinges" proposed by

British consultant engineer A. F. Gimson. According to this design proposal, the three-hinged deck has two hinges on each side of the bridge abutments and one hinge on the top of arch, and the entire load is supported by the two abutments. The steel structure is divided into 13 sections, with a total length of 97.6 meters and the bridge deck width of 19.8 meters, comprising two sideways and a central roadway. The bridge has a clearance of 4.6 meters at the maximum water level. It has a deck gradient of 5%. The bridge abutments are horseshoe-shaped. The steel frame is constructed by riveting curved steel I-beams and steel plates. The foundations of abutments consist of totally 102 wooden piles driven 5.8 *zhang* deep, on top of which the steel beams are fixed. The steel beams weigh 455 tons, and the reinforced concrete on the deck weighs 697 tons, totaling 1,152 tons. The pile driving is done with slopes of 75°, 50° and 17° respectively. The bridge was designed to have a weight-bearing capacity of 20 tons. It was the largest and the most ingenious one-arch bridge in China then. On May 1, 1934, the reconstruction of Ningbo Laojiang Bridge was officially launched.

The preparatory committee paid close attention to the bridge construction work, and the project progressed rapidly. Finally, all the work was completed on May 25, 1936. Since then, a silver gray steel-structured bridge with vermilion railings has been erected on Fenghua River like a rainbow, and Ningbo people's long yearned wish has finally come true. The two Chinese characters "灵桥" (Lingqiao Bridge) written by Tan Yankai, a founding member of the Kuomintang, was hung on top of the east and west sides of the bridge. The steel beams are supported with cement tower-like structures, and their outer walls are steel engravings, including the "重修灵桥碑记" (Record of Rebuilding Lingqiao Bridge) composed by Chen Baolin, with its horizontal board inscribed in seal script by Zhao Shigang and the text written by Sha Wenruo.

After the founding of New China, especially since its reform and opening up, nearly ten bridges have been built across the three rivers. However, no matter how many, how long and how tall the other bridges are, in the eyes of

Ningbo people, Lingqiao Bridge will always be a great old bridge ranking first, a heroic bridge of unyielding endurance. It is the greatest bridge of which Ningbo people are proud!

As the first steel-beam single-arch bridge in China, Lingqiao Bridge embodies the love of "people of the Ningbo Commercial Group" people for their hometown, and has become a symbol of modern Ningbo City as a landmark.

# II. Fangqiao Bridge

Fangqiao Bridge (square bridge) is located in the northeast corner of Fangqiao Village, Jiangkou Sub-district, Fenghua City. It has a critical geographical location because Xianjiang River, Danjiang River and Dongjiang River converged here. Historically, Fangqiao Bridge enjoyed the unique advantages of convenient land and water transportation, as it connected Taizhou and Wenzhou in the south and Ningbo and Shaoxingin the north. It has been situated on the way of the post roads in Eastern Zhejiang since ancient times.

It is said that before the Ming dynasty, there was a small water gate ( 碶 , *qi*) here, commonly known as "Changpu Water Gate". Since it had been in disrepair for a long time, it collapsed under the long-time impact of the river, and therefore a bridge was built and named "Dafang Bridge". It was a simple wooden flat bridge. In the thiry-fifth year of the Qianlong reign (1770), the bridge was taken down and a stone-structured five-arch ring bridge was built and named "Taiping New Bridge". It was honored as "Bridge No. 1 in Eastern Zhejiang". In the twenty-seventh year of the Guangxu reign (1901) , it collapsed. It was not until the thirty-third year of the Guangxu reign (1907) that the current flat bridge with steel beams was built and the name was finalized as "Fangqiao Bridge".

The Fangqiao Bridge is a steel through-type bowstring truss bridge structure, stretching from north to south. It has a total length of 85.5 meters, and a width of 6.02 meters. It has the steel frame of 87.5 tons in total. The bridge deck was paved with wooden boards also designed and constructed by Germans, which was replaced by hollow concrete slabs after 1964. It was constructed nearly 30 years earlier than the Lingqiao Bridge, which also had a steel frame structure. Therefore, a saying goes among the local people that Ningbo Laojiang Bridge is a copy of the Fangqiao Bridge. Unfortunately, in 2007, the first bridge in Ningbo with a long history was damaged after being struck by a cargo ship. Later, after many efforts, in memory of Fangqiao Bridge, a new but smaller bridge was built nearby according to the architectural style of the original one with its preserved components. No transport is allowed on it. And no new bridge is built at the original site.

# III. Huanghunchen Bridge and Yinzhenjiang Bridge

Huanghunchen Bridge (*lit.* dusk-and-dawn bridge) is located in the Renxin Village, Yunlong Town, Yinzhou District. It was built in 1931. It is an arch cement bridge with two piers and three arches, which runs from north to south. The whole bridge is 22.24 meters long, 2.5 meters wide and 4.8 meters high. The bridge entrance is shaped like a dustpan, with the outer opening 5.24 meters wide. The middle arch of the bridge is the largest, spanning 7 meters, and the other two are slightly smaller, spanning 4.5 meters. The east and west sides of the bridge are cement railing panels of 1 meter high. On the outside of the railing panels are inscriptions of some golden rules like "Be filial to your parents, and you shall be blessed; be filial to your parents, and you shall live long; be filial to your parents..." On the central panel are four Chinese characters in regular script: " 黄昏晨桥 " (Huanghunchen Bridge).

To the north of the bridge, there is a cement pavilion with a four-sided pyramidal roof, which is a supporting project of the bridge. Both the bridge and the pavilion are made of reinforced concrete, which are so solid and firm that they have been well preserved so far.

It is said that in gratitude for the kind nurture of his hometown, Mr. Chen Decai from Jiangshan, Yinzhou, who became rich in the construction industry in Shanghai, built ten similar bridges in Ningbo City, among which is Huanghunchen Bridge.

Yinzhenjiang Bridge is located at Jiangqiaotou Village, Xiaogang Sub-district, Beilun District, Ningbo City, and its construction date is unknown. It was originally a five-arch bridge of stone slabs, with wooden railings on one side and a wooden gate on the southeast bank. Under the bridge is torrential river. A precious historical photo of the bridge is still preserved today, in which a young lady dressed in fashionable clothes sat on the bridge railings. It is assumed that she must have been from a family of Jiangqiaotou Village who did business in Shanghai. The bridge has become dangerous due to its long history, the long-time erosion by the tide of Xiaojia River, the disordered ashlars of the piers, the inclined stone slabs, and the lowered position of the west part of the bridge. These conditions are completely consistent with the reasons for the bridge repair recorded on the stone stele.

The present bridge was rebuilt in 1933. It is a reinforced concrete five-arch bridge of European style. The total length is 48.7 meters and the width is 2.68 meters. Yinzhenjiang Bridge crosses Xiaojia River in a northwest-southeast orientation. The bridge features a slightly curved and sloped design. There are two high cement columns at the southeast end of the bridge, which were originally installed to support a gate. The bridge is flanked by openwork cement railings of 0.69 meters high, and the railing panels feature patterns of geometric shapes in a fashionable style. The bridge deck is paved with cement in a grid pattern, which is beautifully simple and can avoid becoming slippery. Three Houjiang Pavilions were built at the northwest end of the bridge, and later were reconstructed into two in 2000. Beside the pavilions, there is a stone stele inscribed with the text entitled "Causes for the Reconstruction of Yinzhenjiang Bridge" composed by Ding Fangyuan, a manager of Baodaxiang Textile Company. Looking from the northwest end to the southeast end, the Guangji Nunnery is on the left. Originally built in the fourth year of the Kangxi reign of the Qing dynasty (1665), it is a typical nunnery architecture in Jiangnan, in which the "Guangji Nunnery Field Stele" preserved has a high value for cultural relics research.

There is another stele nearby entitled "Credit Stele for the Reconstruction of the Yinzhenjiang Bridge", which was set up to make public some information for verification. In addition to the balance sheet, it includes the names of the project proprietor (i.e., project owner), the building contractor (i.e., building company), and the construction supervisor (i.e., project supervisor). As a village of 80 years ago, the project team had an open and advanced mind in that it not only transformed the stone bridge into an advanced cement bridge of European style, but also adopted an advanced construction management form (with the third-party supervision). The middle of the outer sides of the railings bore the regular script " 鄞镇江桥 " (Yinzhenjiang Bridge). Wang Yuxiang, who inscribed the characters, was a famous calligrapher and painter in Ningbo at that time. The piers are designed as thin-walled, light-weight structures, wider at the base and narrower at the top, forming a splayed shape, and the surfaces facing the water are thickened to add strength. There is a cement pole on the northern side of the central arch, which was originally erected as a lamppost. Imagine how glad people might feel who hurried in the dark when they saw the bright light on the bridge! Yinzhenjiang Bridge used to be the boundary bridge between Yin County and Zhenhai. Due to the wide river and the strong current here, the original bridges were repeatedly destroyed and reconstructed. In 1933, businessmen of the origin of Yin County and Zhenhai who had emigrated to Shanghai raised money and built the existent bridge imitating the European style, which was constructed by Shanghai Shenshengji Construction Company at the cost of more than 7,000 silver dollars. After more than 80 years, the bridge remains as solid and stable as before, and is still in good condition.

# Chapter Twenty

# Former Residences of Contemporary Celebrities

Opening the *Dictionary of Tang Poetry Appreciation*, you will find the first poem "Cicada" written by Yu Shinan, a native of Cixi, Ningbo. Yu Shinan (558–638), a poet of the Tang dynasty, was one of the 14 Creditable Officials of Lingyan Pavilion and the director of Palace Library, who died 81 years old. As one of "the Four Greatest Calligraphers in the Early Tang Dynasty" together with Ouyang Xun, Chu Suiliang and Xue Ji, his calligraphy is known to be both forceful and graceful, similar to the style of his poetry, elegant and vigorous.

Yu Shinan is one of the early celebrities in Ningbo, and his former residence has long gone. According to investigations, Yu Shinan's Former Residence had been located on the Site of Dingshui Temple in Minghe Town, Cixi City. In 1998, the government departments set up a stone stele on the site with inscriptions " 唐虞秘监故里 " (Home to Yu Shinan, the director of Palace Library of the Tang dynasty) for commemoration.

Famous people make a famous city. The former residences of celebrities are the precious treasure of Ningbo. However modern a city is, it is superficial if without any historical and cultural heritage.

Ningbo has a galaxy of talents. According to statistics, there were 161 *jinshi* in Ningbo in the Northern Song dynasty, and 983 in the Southern Song dynasty. Among them, Yin County alone witnessed four *zhuangyuan*, ranking first among China's cities. In the Ming dynasty, the number of *jinshi* in Ningbo still ranked first in China.

In modern times, Ningbo has seen an extraordinary surge of talented individuals, and the most famous were people of the Ninbo Commercial Group, well known at home and abroad. For example, Ye Chengzhong, Yan Xinhou and Zhu Baosan, who co-founded the Imperial Bank of China; Kui Yanfang, Liu Hongsheng, Hu Yongqi and Sun Hengfu, who successively founded insurance companies such as Xinping, Dahua, Ningshao, Siming and Tianyi; Yu Qiaqing and Sheng Pihua, who founded the first Chinese stock exchange: Shanghai Stock and Commodity Exchange. Song Hanzhang, one of the main founders of China's modern financial industry, was also a Ningbonese. He was employed as the manager of the Shanghai Branch of the Bank of Great Qing. In 1912, the Bank of Great Qing was reorganized into the Bank of China, which established its nowadays international status.

In addition to the financial industry, Ningbo people also dominated the industries of shipping, hardware, pharmaceuticals and dyes. For example, Yu Qiaqing, a representative of the Ningbo Commercial Group, successively established Sanbei Shipport Company and Ningshao Shipping Company. Other examples are Ye Chengzhong and Zhu Baosan in the hardware industry, Qin Jun'an and Zhou Zongliang in the dye industry, and Huang Chujiu and Xiang Songmao in the pharmaceutical industry. There are also Wu Jintang, Wang Kuancheng, Bao Yugang, Shao Yifu, Ying Changqi, and many others, all of whom are notable for their brilliant achievements.

As native Ningbo people, they left beautiful residences in their hometown Ningbo, which provide historical information for people to understand the Confucian business culture of the Ningbo Commercial Group.

In the long history, many of the former residences of modern celebrities have been exposed to bad weather and man-made destruction, so it is extremely difficult to keep them well preserved. With the accelerating process of urbanization, modern buildings in urban and rural areas are gradually disappearing, including some important former residences of modern celebrities.

Black bricks and white walls, tables, chairs, and many other details of the former celebrities's residences. have been telling their stories of the past. Let's step into their former residences and listen to their stories.

# I. Yu Qiaqing's Former Residence

Among the former residences of modern celebrities in Ningbo, the most exquisite architecture is the old house of Yu's family, Tianxu Hall, a successful example of combining Chinese and Western styles in the architecture.

Yu Qiaqing was a very influential figure in the modern history of China. He had an influence on the development of modern Shanghai and even the politics and economy of the whole China.

Yu Qiaqing was widely involved in modern domestic political activities. Over 60 years from the late Qing to the early period of the Republic of China, he had close contact with the important people of the central and local governments, and his political influence attracted substantial attention. However, he did not seriously consider becoming an official, and always regarded himself as a businessman.

Yu Qiaqing has worked as a comprador, founded many enterprises, and participated in many influential industrial activities, such as Siming Bank, Ningshao Merchant Shipping Company, Sanbei Shipport Company, Sanbei Shipping Group, and so on. Besides, he engaged in extensive social activities. Once as the president of the National Chamber of Commerce and Industry and the president of the Shanghai General Chamber of Commerce, he was enthusiastic in promoting public, good, attaching importance to hometown fellowship, and therefore was a typical leader and icon of the modern Ningbo Commercial Group.

Yu Qiaoqing was extremely filial to his mother, and did everything his mother demanded. After he made his fortune in Shanghai, he had intended to take his mother to Shanghai so that she could have a better life with him, but she was reluctant to part with her hometown, so he built the house for her in the hometown. It was named "Tianxu" with the wishful meaning that his mother would "enjoy talking about the happiness of the family".

Located in Longshan Town, Cixi, Tianxu Hall is 30 kilometers from the urban area of Cixi City and Ningbo City, 1.5 kilometers south of Fulong Mountain, 1.5 kilometers northeast of National Highway G329 from Ningbo to Hangzhou, and 2 kilometers west of the East China Sea in the east. There is a river in front of the house, which is slightly curved and linked to the Citang River at the south end. It goes east to the sea. There used to be busy water transportation with ships coming and going on the river. It has its source in Fengpu Lake of Dapeng Mountain. From the perspective of geomancy, the location of Tianxu Hall can be called "*jincheng huanbao*" (*lit.* surrounded by gold) with "*jin*" (gold) meaning water. In geomancy, the water on the inner side of a bow-shaped river is called "*miangong shui*" (*lit.* sleeping bow-shaped water), which is an auspicious form of water. To the north, Fulong Mountain is like a giant dragon lying on the shore of the East China Sea. It is undoubtedly an "auspicious place", which meets the beliefs of ancient construction that "Keep the wind, you get water" and "Keep the dragon, you gather *qi* together". As far as the requirements of modern living environment are concerned, it is also a desirable place to live in, for there is a river in front of it, which is convenient for boat transportation, washing and fetching water, and plays an important role in regulating the climate and purifying the environment. Besides, Fulong Mountain in the north can function as a block from the biting north wind in winter.

The existing main building has five sections, with a width of 59 meters and a depth of 94 meters, covering an area of 5,546 square meters and a building area of 5,670 square meters. It is separated by a path of 59 meters long and 3.6 meters wide, forming two relatively independent parts at front and back. As a successful example of modern architecture combing Chinese and Western styles, the whole architecture is laid out symmetrically on a central axis. It reflects the shift of the modern Chinese upper class in their architectural aesthetics, and is also the historical epitome of the changes in lifestyle and ideology at that time.

The first three sections at the front part, which were built in 1916 and completed in 1919, is a traditional Chinese wooden structure building, consisting of the screen wall, gate, hall, back building and wing rooms. The screen wall has been destroyed. The gate features a three-bay archway with a splayed facade, with a plaque on the top with four characters of " 天伦乐叙 " (to enjoy talking about the happiness of family life) patterned with smooth-facing bricks. The first section has a lean-to roof, resembling a three-bay hall of *daozuo*. The central bay is connected with the gate. The secondary rooms on the east and west are janitor's rooms, and under the eaves is the round ridge roof. The second section is composed of nine bays and two corridors, including the main hall and the east and west side buildings. In the middle of the main hall, there is a plaque inscribed with three characters " 天叙堂 " (Tianxu Hall). On the beams, tiebeams, corbels and *queti*, there are delicate and beautiful carvings featuring figures from stories, squirrels, Buddha's hand, phoenix and others.

The next two sections were built between 1926 and 1929, only ten years later than the first three sections, but their styles are vastly different. The courtyard has a high gate and an open hall with the wall as high as 6 meters and the main gate in the middle. On the gate is a plaque made of smooth-facing bricks, inscribed with the four characters " 福 禄 欢 喜 " (fortune, wealth, joy and happiness). On the east and west sides of the courtyard are two side gates symmetrically arranged. The plaque on the west side gate is inscribed with the four characters " 增荣益誉 " (to gain more honors and higher reputations).

The plaque and carved patterns on the east side gate have been covered with mud, and therefore the inscriptions are unintelligible. There are also two smaller doors between the main gate and the side gates, and two overpasses were built to connect them with the first three sections. The front of the gate is designed in the traditional style, while the back is similar to the Roman Baroque architectural style, with carved patterns of leaves, curly grass, flowers, and so on. The moldings of the front gate are decorated with strong concave and convex effect in the Western style, which is completely different from the traditional style at the upper part. The main building is a double-eaved two-level building with a gable roof. It has nine bays, two lanes and a front porch, which shows distinctive characteristics of the Western town house. There are fireplaces indoors. The ceiling is generally a dome composed of multi-layer cyma recta. The lattice of doors and windows is patterned with rhombus pieces and tortoise shells. The back building is a double-eaved two-level one with a gable roof.

In terms of construction technology, Tianxu Hall is meticulously crafted with high-quality materials, whether in stonework, brick carving, wood carving or concrete structures. In particular, the concrete structures and decorations, despite decades of erosion, show minimal signs of cracking, crumbling or peeling. The mosaic floor and ceramic tiles on walls are still intact and bright in color. The lines of the concrete cornices are clear-cut. Patterns of leaves, curly grass, hanging curtain, geometric figures and other decorations made of concrete at the upper part of the colonnades and walls are neat and exquisite.

The tiebeams, corbels, *queti*, lintels, couplets, columns and some other parts are decorated with intricately carved patterns of phoenix, peonies, deer holding ganoderma lucidum between its teeth, the Three Friends in Cold Weather, the Four Gentlemen, antiques, apricot flowers, crabapples, and scenes from the *Romance of the Three Kingdoms* and *Journey to the West*. The whole building complex looks gorgeous and noble and shows high artistic value.

# II. Former Residence of Chiang Kai-shek

Among the residences of former celebrities in Ningbo, the Former Residence of Chiang Kai-shek is the best known, located on Wuling Road, Xikou Township, Fenghua City. It used to be resided by Chiang's family, including Chiang Kai-shek and his father. Built between the 1930s and 1940s, it consists of Fenghao House, Western-style House and Yutai Salt Shop, among others. The building complex is a combination of Chinese and Western styles, with a unique structure, so much so that it became a valuable historic spot of the period of the Republic of China and a representative architecture of typical Jiangnan Residence.

"Wearing military uniforms, I patrol the mountains in my hometown. / The heavily guarded villa on top of the hill is the place I dwell. / I, the owner of the villa, came up with a brilliant plan to build a peaceful world, / And hardly had I blurted out 'Great!' when the ground began to shake." The poem "Miaogao Terrace" describes the villa of the combined Chinese and Western styles built by Chiang Kai-shek in 1930, which has a total construction area of 436 square meters.

According to the sixth volume of *Chiang's Genealogy in Wuling* recompiled in 1949, Chiang Kai-shek was eight years old when he "first ascended Xuedou Mountain and was fascinated by Miaocen Ridge". The villa displays a combination of Chinese and Western styles, with one bungalow on both sides of the gate, both having a flat roof terrace. There are two-story and three-bay buildings behind the courtyard. The cement corridors on the upper floor of the buildings are connected with the roof terraces. Further inside, the middle gate is equipped with a plaque with three black characters " 妙高台 " (Miaogao Terrace) written on the white background by Chiang Kai-shek. There is a three-room bungalow behind. The walls are connected into one, surrounded by green grass and trees. The villa has the view of beautiful mountains and river. It is said that Chiang Kai-shek had been here before fleeing the mainland in May 1949. Looking at the view, he couldn't help feeling sad because the scenery remained the same, but his power has gone. At that moment, he was no longer thinking about his national and military affairs. The familiar local accent and the flowing Fenghua River evoked nothing but the bleakness of his own tragic ending. There is Fuhu Cave on the left, Xiaofan Terrace on the right and Yanzuo Terrace in front of Miaogao Terrace, all of which are relics of the eminent Monk Zhihe of the Song dynasty. It is said that Monk Zhihe began his lecture at Yanzuotai at 5:00 am every day. And a legend goes that there were two tigers in Fuhu Cave listening to his chanting all the year round, so that their wildness gradually faded. Sadly, Yanzuo Terrace was destroyed in the autumn of 1968 and reconstructed with the fund allocated by the state in 1987.

Fenghao House, which was named after Haojing, the capital during the reign of King Wu of Zhou, is located at No. 77, Wuling West Road, Xikou Town. During the reign of Guangxu of the Qing dynasty, there were three old-style buildings. The front building and the second building both consist of seven bays and two lanes. The back building had three bays. The central hall was the front main hall. In the fourteenth year of the Guangxu reign of the Qing dynasty (1888) when Chiang Kai-shek was two years old, because of the fire of the Yutai Salt Shop, the whole family moved to the west wing of Baoben Hall. In the twenty-first year of the Guangxu reign of the Qing dynasty (1895), Chiang Kai shek's father, Chiang Su'an died, and the sons divided up the family property and began to live apart the next year, with Chiang Kai-shek and his younger brother Chiang Jui-ching inheriting Fenghao House. In the twenty-third year of the Guangxu reign of the Qing dynasty (1897), Chiang died, leaving Fenghao House to Chiang Kai-shek alone.

After the victory of the Northern Expedition, Chiang Kai-shek planned to expand his ancestral house. Starting from 1932, it was expanded multiple times and completed in 1935. Fenghao House covers an area of 4,800 square meters, with a building area of 1,850 square meters. It shows a traditional Jiangnan style typical of aristocratic families, compromising a front hall, a back hall, two wing rooms and four corridors, with buildings and pavilions connected with each other, and colonnades attached to the buildings around the courtyard. From south to north, there are 49 rooms, large and small, including the main gate, the front courtyard, the inner gate, the front hall, the inner courtyard, the back hall (i.e., Baoben Hall), the east and west wing rooms, the east building, the former residence of Chiang Kai-shek's mother, and the west bungalow. Facing the street, the gate has an antique gable roof, under which is a plaque inscribed with "Chiang's Former Residence" written by Sha Menghai in 1989. There are also a plaque of "Fenghao House" on the inner gate, and a plaque with the inscription of " 素居 " (*lit.* simple house) on the gate of the front hall. Originally built as the front main hall with three rooms, Baoben Hall was later rebuilt by Chiang Kai-shek. It has a gable roof, gable walls, and roof ridges adorned with sculptures of "fortune, wealth and longevity", "double dragons playing with a pearl", and "phoenix ready to take off". There is a horizontal plaque with gold characters on a red background in the hall corridor, which reads " 寓理帅气 " (embracing reason and commanding spirit). It was written by Chiang Kai-shek on April 12, 1949 on the 40th birthday of his eldest son Chiang Ching-kuo. The two columns in the hall are engraved with couplets: "It is the highest moral principle to repay ancestors and respect parents, and it is filial piety to glorify the older generations and benefit the future ones." The eave columns are carved with scenes from *The Romance of the Three Kingdoms*, and such historical stories of the Western Zhou dynasty as "Fishing in the Weishui River", "King Wen of Zhou Hauling the Coach", "Prince Seeking for Talents", "Battle with Ma Chao at Night", "Guan Yu's Battle in Changsha", and "Guan Yu Back to Jingzhou". The plaque with inscriptions " 报本堂 " (Baoben Hall) in the central bay was written by Wu Jingheng. The carvings of the wing room pillars are based on Yue Fei's (a Chinese military general during the Southern Song dynasty and national hero of China) stories such as "The Flood in Tangyin", "Studying under Zhou Tong", "Yue Fei's Tattooing Characters on His Back" in *Biography of the Patriot Yue Fei* and the stories of the Three Kingdoms.

Chiang Ching-kuo's Western-style House is located to the east of Wenchang Pavilion, close to Shanxi River. It comprises three Western-style buildings with flat roofs. Its building was funded by Chiang Kai-shek in 1930. It covers an area of about 300 square meters. On the roof balcony, which is surrounded by cement railings, one can enjoy the cool and the moon. This small Western-style house is as well designed as Wenchang Pavilion.

On April 27, 1937, Chiang Ching-kuo returned to Xikou from the Soviet Union with his wife Chiang Fang-liang and his son Chiang Hsiao-wen. He lived and studied here until September when he went to Xiong Shihui's office in Jiangxi. The east room upstairs is the bedroom of Chiang Ching-kuo and his wife, and the west room is his study. In the middle is the reception room. In order for Chiang Ching-kuo and his wife to learn Chinese, Chiang Kai-shek invited a teacher named Xu Daolin, the son of Xu Shuzheng, a minister of the Beiyang Army. Xu Daolin, who was then the secretary of the Nanchang Military Camp Design Committee in Jiangxi Province, who was highly respected by Chiang Kai-shek. Xu Daolin, his Italian wife and a housemaid lived downstairs in the Western-style House. Chiang Kai-shek also sent for his distant relative, a female teacher, from Cixi to teach Chiang Fang-liang Chinese, whose Chinese name was given by Chiang Kai-shek.

In December 1939, to mourn the unexpected death of Mao Fumei, his mother, Chiang Ching-kuo wrote four characters " 以血洗血 " (blood for blood) and had it engraved on a stone tablet, which is now kept downstairs. Many important figures of the Republic of China lived in the Western-style House, such as Chen Bulei, who wrote *Half-month Records in Xi'an* there.

# III. Weng Wenhao's Former Residence

Weng Wenhao (1889–1971) is one of the few celebrities of the period of the Republic of China with two residences in Ningbo. As one of the founders of modern geology, geography and seismology in China, he compiled the first colored geological map of China, *The Geological Survey of China*, and was the first Chinese scholar to introduce Alfred Wegener's theory of continental drift. He served as the President of the Executive Yuan of the National Government. After the founding of New China, he was elected a member of the Chinese People's Political Consultative Conference. Mao Zedong referred to Weng Wenhao as a "patriotic Kuomintang military and political person" in *On the Ten Major Relationships*.

One of his former residences is located at Shitang Village, Gaoqiao Town, Yinzhou District, and the other is located at No. 11, Dashuyuan Lane, Haishu District. Both are among the Historical and Cultural Sites Protected at the Provincial Level.

The residence at Shitang Village, the birthplace of Weng Wenhao, was originally built by his great grandfather Weng Jinghe in the Qing dynasty. The Weng's family built up the family fortunes by dealing in Western-style cloth and Weng Jinghe's generation was the family's heyday when their property reached more than 2 million silver taels. As a residence of the Ming and Qing architectural style, it covers an area of 400 square meters, with two two-story buildings at the front and back. Weng Wenhao had fond memories of Shitang Village and his former residence in his old age, as reflected in the following poem.

The beautiful Shitang Village in the west of Yinzhou District is located near the river and against the mountains.

The water for the irrigation of rice is kept by the bridge and the water gate, and the lovely scenery can be seen from the window.

There are plaques to honor brothers in Weng's family for their success in *keju* examination.

There are also memories of having good time with the neighboring farmers.

They taught me personally when I was a beginner, for which I was deeply grateful.

When Weng Wenhao was eight years old, his residence in Shitang Village was robbed by outlaws, which caused panic in the whole family. As a result, Weng's family moved to Yinxian Bridge on the North Riverbank, and then moved to a triangular plot area in front of Tianfeng Temple. Finally, in 1907, they built a new house in Dashuyuan Lane on the west side of Moon Lake. This small building with a *sanheyuan* has a Roman-style brick and concrete structure. The main building faces east, with the main gate facing north. The two sides of the gate are carved with patterns of flowers and plants, the stone architrave carved with curly grass patterns, and the two columns decorated with brick carvings of vase patterns. The well-preserved residence is among the few outstanding buildings of the Republic of China in Ningbo.

There are many other former residences of Ningbo modern celebrities, such as those of Bao Yugang and Wu Jintang. The former residences of celebrities exist as the cultural heritage of the city, offering important information for the study of the historical development of the city. Some former residences also have a high artistic value, which provides significant humanity resources. Those spaces, beautiful or ordinary, spacious or small, leave behind the ethos and the aesthetics of the times. The rich heritage of celebrities' former residences in Ningbo provides an important platform for us to promote the culture represented by them and enhance the ethos of the city.

Bibliography

Chen, H. X. (ed.). *A Surging Tide in the North of Ningbo City: Research on the Old Bund of Ningbo in Modern Times.* Ningbo: Ningbo Publishing House, 2008. [ 陈宏雄，主编 . 潮涌城北——近代宁波外滩研究 . 宁波：宁波出版社，2008.]

Chen, Z. H. *A History of World Architecture: Before the End of 19th Century.* Beijing: China Architecture & Building Press, 1979. [ 陈志华 . 外国建筑史（十九世纪末叶以前）. 北京：中国建筑工业出版社，1979. ]

Chen, Z. H. *Twenty Lessons on Ancient World Architecture (Collector's Edition) (Illustrated).* Shanghai: SDX Joint Publishing Company, 2001. [ 陈志华 . 外国古建筑二十讲：插图珍藏本 . 北京：生活·读书·新知三联书店，2001.]

Dong, Y. A. (ed.). *A Collection of Cultural Relics in Ningbo.* Beijing: Huaxia Publishhing House, 1996. [ 董贻安，主编 . 宁波文物集粹 . 北京：华夏出版社，1996.]

Hong, T. C. *Dongyang's Residences of the Ming and Qing Dynasties.* Shanghai: Tongji University Press, 2000. [ 洪铁城 . 东阳明清住宅 . 上海：同济大学出版社，2000.]

Jin, P. S. & Sun, S. G. (eds.). *Comprehensive Dictionary of Businessmen of Ningbo Origin.* Ningbo: Ningbo Publishing House, 2001. [ 金普森，孙善根，主编 . 宁波帮大辞典 . 宁波：宁波出版社，2001.]

Le, C. Y. *The Outline of Ningbo History.* Ningbo: Ningbo Publishing House, 1995. [ 乐承耀 . 宁波古代史纲 . 宁波：宁波出版社，1995.]

Lin, S. M. *Changes in Ningbo: A Historical Account of Ningbo's Urban Development.* Ningbo: Ningbo Publishing House, 2002. [ 林士民 . 三江变迁——宁波城市发展史话 . 宁波：宁波出版社，2002.]

Lin, S. M. & Shen, J. G. *The Great Silk Road: Ningbo and Marine Silk Road.* Ningbo: Ningbo Publishing House, 2002. [ 林士民，沈建国 . 万里丝路——宁波与海上丝绸之路 . 宁波：宁波出版社，2002.]

Liu, D. Z. (ed.). *Ancient Chinese Architecture History.* Beijing: China Architecture & Building Press, 1980. [ 刘敦桢，主编 . 中国古代建筑史 . 北京：中国建筑工业出版社，1980.]

Liu, Y. *An Outline of the Comparative Study on Chinese and Western Architectural Aesthetics*. Shanghai: Fudan University Press, 2008. [ 刘月 . 中西建筑美学比较论纲 . 上海：复旦大学出版社，2008.]

Lou, Q. X. *Twenty Lessons on the Architecture of Ancient China*. Shanghai: SDX Joint Publishing Company, 2004. [ 楼庆西 . 中国古建筑二十讲 . 北京：生活·读书·新知三联书店，2004.]

Ningbo Archaeological Institute of Cultural Relics. (ed.). *A Collection of Archaeological Studies on Ningbo Cultural Relics*. Beijing: Science Press, 2008. [ 宁波市文物考古研究所，编 . 宁波文物考古研究文集 . 北京：科学出版社，2008.]

Ningbo Protection and Management Institute of Cultural Relics, et al. (eds.). *Ningbo & the Marine Silk Road*. Beijing: Science Press, 2006. [ 宁波文物保护管理所，等编 . 宁波与海上丝绸之路 . 北京：科学出版社，2006.]

Pan, G. X. (ed.). *A History of Chinese Architecture* (6th Edition). Beijing: China Architecture & Building Press, 2009. [ 潘谷西，主编 . 中国建筑史 . 6 版 . 北京：中国建筑工业出版社，2009.]

Qi, S. T. (ed.). *Foreign Affairs in Their Entirety: The Daoguang Reign*. Beijing: Zhonghua Book Company, 1964. [ 齐思和，整理 . 筹办夷务始末（道光朝）. 北京：中华书局，1964.]

Qian, M. W. (ed.). *The History and Traditional Culture of Ningbo*. Ningbo: Ningbo Publishing House, 2007. [ 钱茂伟，编著 . 宁波历史与传统文化 . 宁波：宁波出版社，2007.]

Wang, M. M., et al. *The General History of Ningbo: Republic of China*. Ningbo: Ningbo Publishing House, 2009. [ 王慕民，等 . 宁波通史（民国卷）. 宁波：宁波出版社，2009.]

Wang, T. Y. (ed.). *A Collection of Old Treaties Between China and Foreign Countries*. Shanghai: Shanghai University of Finance and Economics Press, 2019. [ 王铁崖，编著 . 中外旧约章汇编 . 上海：上海财经大学出版社，2019.]

Wang, X. Y. (ed.). *Ethereal Echoes of the Heaven: Religious Architecture*. Beijing: China Architecture & Building Press, 2011. [ 王谢燕，编著 . 飘

渺余蕴天国境：宗教建筑 . 北京：中国建筑工业出版社，2011.]

Xu, Z. B. *Siming Topics*. Ningbo: Ningbo Publishing House, 2000. [ 徐兆昺 . 四明谈助 . 宁波：宁波出版社，2000.]

Yang, F. Y. (ed.). *The Bund Culture and the City Development*. Shanghai: Shanghai Far East Publishers, 2004. [ 杨馥源，主编 . 外滩文化与城市发展 . 上海：上海远东出版社，2004.]

Yu, F. H. (ed.). *The Annals of Ningbo City*. Beijing: Zhonghua Book Company, 1995. [ 俞福海，主编 . 宁波市志 . 北京：中华书局，1995.]

Yu, F. H. (ed.). *The Annals of Ningbo City (expanded)*. Beijing: Zhonghua Book Company, 1998. [ 俞福海，主编 . 宁波市志外编 . 北京：中华书局，1998.]

Yu, P. L. & Ying, K. J. *The Ancient Opera Stages in Ninghai*. Beijing: Zhonghua Book Company, 2007. [ 徐培良，应可军 . 宁海古戏台 . 北京：中华书局，2007.]

Zhang, F. H. (ed.). *Study and Preservation of Chinese Modern Architecture*. Nanjing: Southeast University Press, 2006. [ 张复合，主编 . 中国近代建筑研究与保护 . 北京：清华大学出版社，2006.]

Zhang, G. Q. (ed.). *The Collection of Mingzhou Stele Inscriptions in Tianyige Museum*. Shanghai: Shanghai Classics Publishing House, 2008. [ 章国庆，编著 . 天一阁明州碑林集录 . 上海：上海古籍出版社，2008.]

Zhang, G. Q. & Qiu, Y. P. (eds.). *The Annals of Surviving Steles of All Dynasties in Ningbo*. Ningbo: Ningbo Publishing House, 2009. [ 章国庆，裘燕萍，编著 . 甬城现存历代碑碣志 . 宁波：宁波出版社，2009.]

Zhang, S. Q. *Pictures of Five Mountains and Ten Temples and Architectures of Zen Temples in Jiangnan*. Nanjing: Southeast University Press, 2000. [ 张十庆 . 五山十刹图与江南禅寺建筑 . 南京：东南大学出版社，2000.]

Zheng, S. C. *The History of Ningbo Port*. Beijing: China Communication Press, 1989. [ 郑绍昌 . 宁波港史 . 北京：人民交通出版社，1989.]

Zheng, X. X. *The Preservation of Memory of the Architectural Culture in China*. Beijing: China Architecture & Building Press, 2007. [ 郑孝燮 . 留住我国建筑文化的记忆 . 北京：中国建筑工业出版社，2007.]

Zhou, S. F. *Old Ningbo Culture Series: Old city of Ningbo*. Ningbo: Ningbo Publishing House, 2008. [周时奋. 宁波老城. 宁波：宁波出版社，2008.]

Zhou, S. F. & Xiang, D. *Old Ningbo Culture Series: Old qiangmen in Ningbo*. Ningbo: Ningbo Publishing House, 2008. [周时奋，相栋. 宁波老墙门. 宁波：宁波出版社，2008.]

Glossary

| | |
|---|---|
| 1911 Revolution | 辛亥革命 |
| a change of scenery in each step of the journey | 移步换景 |
| *A Dream of Red Mansions* | 《红楼梦》 |
| *A Field Survey of China's Export Trade* | 《中国出口贸易实地考察》 |
| a five-disk phase wheel | 五重相轮 |
| A. Corrit Co. | 康益洋行 |
| A Note on Cixi County School | 慈溪县学记 |
| Abolishing the Imperial Examination and Practicing New Schooling | 废科举，行新学 |
| academy | 书院 |
| *anceng* (*lit.* dark stories, or those that are not marked on the outside) | 暗层 |
| ancestral hall | 祠堂 |
| Anlan Guild Hall | 安澜会馆 |
| An-Shi Rebellion | 安史之乱 |
| arcade style | 券廊式 |
| arch bridge | 拱桥 |
| arch structure | 拱券结构 |
| Archaeological Society of China | 中国考古学会 |
| architectural complex of the Ming dynasty in Cicheng | 慈城明代建筑群 |
| architrave | 额枋 |
| Arhat Hall | 罗汉殿 |
| Ashoka Temple | 阿育王寺 |
| Associate Administrator | 同知 |
| auspicious place | 吉地 |
| Bailiang Bridge (*lit.* bridge of hundreds of beams) | 百梁桥 |
| Baiyun Manor | 白云庄 |
| balustrade capitals | 栏杆柱头 |
| balustrade panel | 栏板 |

Bamin Guild Hall ("Bamin" is Fujian; "ba", literally "eight", represents eight administrative divisions of Fujian Province; "Min" is the short form for Fujian Province) 八闽会馆

*bangyan* (a scholar who ranked second in the highest imperial examination) 榜眼

Banpu 半浦

Baoguo Temple 保国寺

*baogushi* (drum-shaped gate bearing stone pillar) 抱鼓石

*baotou* beam (*lit.* beam holding the head, or the beam connecting an eave column and an interior column) 抱头梁

barrel tile 筒瓦

bay 开间

beam-column style 梁柱式

Bessemer steel making process (converter steel making) 酸性转炉炼钢

*Biography of the Patriot Yue Fei* 《精忠岳传》

black rock 青石

block 斗

bone spade 骨耜

borrowed scenery 借景

boxwood carving 黄杨木雕

bracket arm 拱

bracket set 斗栱

branch ancestral hall 支祠

brick carved palace gate 砖雕宫门

brick-making technology 砖石结构

bridge pier 桥墩

British Merchant Firm 英商洋行

| | |
|---|---|
| Buddhist relics | 舍利 |
| building component | 建筑构件 |
| *Building Standards* (*Yingzao Fashi*) | 《营造法式》 |
| Bureau of Works of Shanghai International Settlement | 上海公共租界工部局 |
| buttress | 扶壁 |
| *caifangshi* (an official monitoring imprisonment and provincial and county government officials) | 采访使 |
| caisson ceiling | 藻井 |
| *caotai* stage (the stage that is made temporarily with anything possible, like several tables put together) | 草台 |
| carved board | 花板 |
| Catholic Abbey | 天主教堂修道院 |
| *cejiao* (approximately 1% slight inward incline of the column on the top) | 侧脚 |
| celadon porcelain | 青瓷 |
| central bay | 明间 |
| *cha* (the top of stupa) | 刹 |
| *chagong* (bracket-arm directly inserted into another member such as the column shaft) | 插栱 |
| *chandu chuomu* | 蝉肚绰幕 |
| Changping Barn (the barn to stabilize the grain price) | 常平仓 |
| Changpu Water Gate | 常浦碶 |
| Chapter "Biography of Xiang Minzhong" in *The History of the Song Dynasty* | 《宋史·向敏中传》 |
| Chapter "Documentary Annals" in *The General Annals of Yin County* | 《鄞县通志·文献志》 |
| Chapter "Duke Wen of Teng II" in *Mencius* | 《孟子·滕文公下》 |
| Chapter "Explaining the Hexagrams" in *The Book of Changes* | 《周易·说卦》 |

| | |
|---|---|
| Chapter "Patriarchal Law" in *The Inquiry into the Rituals* | 《学礼置疑·宗法》 |
| Chapter "Yongye" in *The Analects* | 《论语·雍也》 |
| *chashou* (inverted V-shaped brace) | 叉手 |
| *chengchen* (the ceiling that serves to block the dust from the eaves) | 盛尘 |
| *Chiang's Genealogy in Wuling* | 《武岭蒋氏宗谱》 |
| Chief Deputy Envoy to Liao | 辽主副使 |
| Chief Procurator | 检察长 |
| China Engineer Co. | 信昌洋行 |
| Chinese Academy of Engineering | 中国工程院 |
| Chinese People's War of Resistance Against Japanese Aggression | 抗日战争 |
| Chinese regular script | 正书 |
| *chitou* (the component at the upper part of the gable jutting out of the gable to support the eaves) | 墀头 |
| *chiwei* (owl's-tail-shaped ornament at the end of a roof ridge) | 鸱尾 |
| "Cicada" | 《蝉》 |
| Chogen | 重源 |
| Chongde Yishu (Chongde Free Private School) | 崇德义塾 |
| *chongqi* | 重棋 |
| Chongsheng Temple | 崇圣祠 |
| *chuantai* stage (the stage that is assembled temporarily with separate wooden parts) | 串台 |
| *chujiang* (outgoing of a general) | 出将 |
| *chuomu* tie-beam | 绰幕枋 |
| cicada-belly-shaped *queti* | 蝉肚式雀替 |
| *Cihai* (*Chinese Dictionary of Etymology*) | 《辞海》 |
| Cihu Middle School | 慈湖中学 |

| | |
|---|---|
| Citang Bridge | 祠堂桥 |
| City God | 城隍神 |
| City God Temple | 城隍庙 |
| clan temple | 家庙 |
| clearing house | 票据交换所 |
| clerical script | 隶书 |
| coiling dragon | 盘龙 |
| colonial-style architecture | 殖民地式建筑 |
| column base | 柱础 |
| column-and-tie-beam framework | 穿斗式结构 |
| column-beam-and-strut framework | 抬梁式结构 |
| column-top bracket sets | 柱头科 |
| commander-in-chief | 都督 |
| *Complete Collection of Pictures and Books of Old and Modern Times* | 《古今图书集成》 |
| *Complete Library of the Four Treasuries* | 《四库全书》 |
| *Comprehensive Dictionary of Businessmen of Ningbo Origin* | 《宁波帮大辞典》 |
| conferment of honorary titles for three generations by imperial mandate | 诰封三代 |
| Confucian culture | 儒家文化 |
| Confucian scarf (a sort of headwear) | 儒巾 |
| Confucius Temple in Qufu | 曲阜祖庙 |
| Confucius Temple | 孔庙 |
| Confucius Temple School | 孔庙学宫 |
| construction supervisor (i.e., project supervisor) | 监理方 |
| corbelled cornices | 叠涩出檐 |
| Corinthian column | 科林斯罗马柱 |
| countrymen guild hall | 同乡会馆 |

| | |
|---|---|
| county magistrate | 县令 |
| county school | 县学 |
| county-level imperial examination | 乡试 |
| Court of Censors | 都察院 |
| courtesy name | 字 |
| credit stele | 征信碑 |
| cross beam | 梁架 |
| cross-shaped bracket set | 十字科 |
| cultural layer | 文化层 |
| curled-up dragon stone column | 蟠龙石柱 |
| *Customs and Traditions* | 《风俗通》 |
| Dacheng Gate | 大成门 |
| Dacheng Hall | 大成殿 |
| Dafang Yuedi Residence | 大方岳第 |
| Daihongzan Eiheiji Temple | 大本山永平寺 |
| *daogua* building | 倒挂楼 |
| *daotai* (circuit intendant, supervisor of special administration zones directly subordinated to central government agencies) | 道台 |
| *daowu* (the south room) | 倒屋 |
| *daozuo* | 倒座 |
| Dayin rock | 大隐石 |
| Deputy Commissioner of the Surveillance Commission | 按察司副使 |
| Dhanari column | 经幢 |
| *Dictionary of Tang Poetry Appreciation* | 《唐诗鉴赏辞典》 |
| diminishing | 递减 |
| director of Palace Library | 秘书监 |
| disc-shaped column base | 櫍形柱础 |
| District Cultural Administration Office | 区文管办 |

| | |
|---|---|
| District Magistrate | 知县 |
| District-owned Jingye Combined Primary School | 区立敬业完全小学 |
| Dongming Cottage | 东明草堂 |
| Dongqian Lake | 东钱湖 |
| Dongxiaozi Temple | 董孝子祠 |
| *dongxiwu* (the side halls on the east and west of Dacheng Hall) | 东西庑 |
| double-side columned veranda | 两面列柱外廊 |
| *douqi* | 斗棋 |
| dovetail joints | 燕尾榫 |
| dragon vein (representing the path of *qi* flow) | 龙脉 |
| dragon-phoenix stone column | 龙凤石柱 |
| Drum Tower | 鼓楼 |
| drum-shaped | 鼓式 |
| Duke of Wenxuan, the Supreme Sage of Great Accomplishment | 大成至圣文宣王 |
| dukes and princes | 诸侯 |
| Dunshan Hall stage | 敦善堂戏台 |
| dwarf column | 童柱 |
| East Asian cultural circle | 东亚文化圈 |
| East Well | 东井 |
| Eastern Zhejiang School | 浙东学派 |
| eave purlin | 檐檩，正心桁 |
| eaves-end tile | 瓦当 |
| eclecticism | 折中主义 |
| eight heavenly horses | 八天马 |
| Eight Immortals table (a square table for eight people) | 八仙桌 |
| eight slightly-cut angles | 小八抹角 |
| Eisai | 荣西 |

| | |
|---|---|
| Emperor Gaozu of the Han dynasty | 汉高祖 |
| Emperor Qianlong | 乾隆帝 |
| Engineering Bureau | 工程局 |
| entasis | 卷杀 |
| "Erling Mountain" | 《二灵山》 |
| *Essays in Suiyuan* | 《随园随笔》 |
| examination hall | 考棚 |
| Executive Yuan Administrative Ministry of the Nationalist Government of the Republic of China | 国民政府行政院 |
| Famed Literature Land | 文献名邦 |
| Fan Qin | 范钦 |
| Fangqiao Bridge (square bridge) | 方桥 |
| Fan's Residence | 范宅 |
| Feng Yue's painted *taimen* | 冯岳彩绘台门 |
| Fengguo Junlou Shrine (the memorial temple in honor of the dedicated army) | 奉国军楼神祠 |
| Fenghao House | 丰镐房 |
| fingered citrons (alias Buddha's hand, as the fruit looks like the Buddha's hand; "佛" is the homonym of "福") | 佛手 |
| fire-resistance gable wall | 封火山墙 |
| five bats holding a longevity peach (symbolizing happiness and longevity) | 五蝠捧寿 |
| flagpole pedestal | 夹杆石 |
| flat tile | 板瓦 |
| flying rafter | 飞椽 |
| *Foreign Affairs in Their Entirety* | 《筹办夷务始末》 |
| foreign dragon | 洋龙 |
| Forest of Steles in Mingzhou | 明州碑林 |
| Former Residence of Chiang Kai-shek | 蒋氏故居 |

| | |
|---|---|
| *fotu* (Buddha's Figure) | 佛图 |
| foundation | 基址 |
| four styles of calligraphy (regular script, cursive script, clerical script, seal script) | 书法四体（正书，草书，隶书，篆书） |
| four treasures of the study (ink, inkstone, mountain-shaped pen holder and white paper) | 文房四宝 |
| four-side columned veranda | 四面列柱外廊 |
| Frederick Academy | 斐迪书院 |
| front overhanging eave | 前出檐 |
| Fu Xinian | 傅熹年 |
| Fufu Room (*lit.* the room of curled-up insteps) | 伏跗室 |
| Fujiashan Site | 傅家山遗址 |
| *fuma* (emperor's son-in-law) | 驸马 |
| Fuma Mo's Former Residence | 莫驸马宅 |
| *fuxue* | 府学 |
| *fuzi mentou* (*lit.* the gate with Chinese character "福", which means blessings) | 福字门头 |
| gable roof | 硬山顶 |
| gable-and-hip roof | 歇山顶 |
| *gaomen* | 皋门 |
| gatehouse | 门楼 |
| General Surveillance Circuit | 巡道 |
| General Wuji (Courageous Cavalryman) | 武骑将军 |
| *German Businessmen in Ningbo* | 《德商甬报》 |
| girder bridge | 梁桥 |
| God of land | 土地神 |
| God of wealth | 财神 |
| gold foil decal | 金箔贴花 |
| *gongsuo* (guild office) | 公所 |

| | |
|---|---|
| Goujian | 勾践 |
| *goulian daqian* structure (connected beam frame structure) | 勾连搭牵梁架结构 |
| Gouzhang | 句章 |
| government office | 官署 |
| Grand Canal | 大运河 |
| Grand Master for Forthright Service | 奉直大夫 |
| Grand Mentor of the crown prince | 太子太傅 |
| Grand Secretariat | 内阁 |
| Grand Secretaries | 阁老 |
| graveyard ancestral temple | 墓祠 |
| Great Hall of Baoguo Temple | 保国寺大殿 |
| ground plan | 平面形制 |
| *gualeng* column (melon-shaped column) | 瓜棱柱 |
| *gualuo* (hanging fascia board) | 挂落 |
| Guangji Bridge | 广济桥 |
| Guangji Nunnery Field Stele | 广济庵田碑 |
| *guanyindou* style (a style of the gable resembling the hood of Guanyin's Bodhisattva) | 观音兜 |
| "Guidelines for Zhongtian Pavilion Disciples" | 《中天阁勉诸生》 |
| *guijiao suyun* | 圭角素云 |
| *gumen* (ancient gate) | 古门 |
| *guqin* (a plucked seven-string Chinese musical instrument) | 古琴 |
| Haishu District | 海曙区 |
| *Half-month Records in Xi'an* | 《西安半月记》 |
| Hall of Guanyin Bodhisattva | 观音殿 |
| Hall of the Four Heavenly Kings | 天王殿 |
| Han people | 汉人 |

| | |
|---|---|
| Hanlin Academy | 翰林院 |
| harmony between nature and humanity | 天人合一 |
| Headquarters of Ningbo Prefecture | 宁波府正堂 |
| Heaven is one, and it gives birth to water; earth is six, and it gives shape to it. | 天一生水，地六成之。 |
| Hefeng Yarn Factory | 和丰纱厂 |
| Hemudu Culture | 河姆渡文化 |
| Hemudu Site | 河姆渡遗址 |
| Hengsheng Village | 横省村 |
| high relief | 高浮雕 |
| hip roof | 庑殿顶 |
| hollow plain wall | 空斗清水墙 |
| hollowed-out brick window | 砖漏窗 |
| *huagong* (any bracket arm in the bracketing structure that projects away from the wall) | 华栱 |
| *hualan* beam (cross-shaped beam) | 花篮梁 |
| Huang Zongxi | 黄宗羲 |
| Huanggong Ferry | 黄公渡 |
| *hudou* (the supporting bracket) | 护斗 |
| Huguang Province | 湖广 |
| *huibei* (an adhesive substance to stick tiles together) | 灰背 |
| Huide Bridge | 惠德桥 |
| *huiguan* (guild hall) | 会馆 |
| *huiyuan* (a scholar who ranked first in the metropolitan examination) | 会元 |
| Huizhou architecture | 徽派建筑 |
| imperial court edition | 内府本 |
| imperial edict | 圣旨 |
| imperial examination | 科举 |

| | |
|---|---|
| "In the Mountains" | 《山中》 |
| industry guild hall | 行业会馆 |
| intaglio carving | 沉雕，阴雕 |
| intercolumnar bracket sets | 平身科（斗栱） |
| intermediate purlin | 金坊 |
| interspersed style | 穿套式 |
| Jean-Alexis de Gollet | 郭中传 |
| Jiadi Shijia Residence | 甲第世家 |
| Jiang Jingguo's Western-style House | 蒋经国小洋房 |
| Jiangbei Catholic Church | 江北天主教堂 |
| Jiangbei Police Station | 江北巡捕房 |
| Jiangnan (lower reaches of the Yangtze River) | 江南 |
| *jiaoshouguan* (professor in charge of education) | 教授官 |
| *jiatang* board | 夹堂板 |
| *jibu* | 脊步 |
| jielü jiehua zhuang (red paintings with blue and green margins on architectural components) | 解绿结华装 |
| *jigua* column | 脊瓜柱 |
| *jilongding* (the type of spiral caisson ceiling that looks like the top of a chicken cage) | 鸡笼顶 |
| *jinbu* | 金步 |
| *Jing* (classics), *Shi* (history), *Zi* (philosophy), *Ji* (literature) | 经，史，子，集 |
| Jingye Primary School for Senior Classes | 敬业高等小学堂 |
| Jingyi Pavilion | 敬一亭 |
| *jinshen* (government officials) | 缙绅 |
| *jinshi* (a scholar who passed the highest imperial examination) | 进士 |
| John Fitzgerald Brenan | 毕约翰 |

| | |
|---|---|
| *Journey to the West* | 《西游记》 |
| *juren* (successful candidate in the imperial examinations at the provincial level) | 举人 |
| *jushu* (the vertical rise of the roof) | 举数 |
| Juxinghui Association | 聚姓会 |
| Kan Ze | 阚泽 |
| King Wu of Zhou | 周武王 |
| Korakuen Garden | 后乐园 |
| Kuaiji Prefecture | 会稽郡 |
| Kuomintang | 国民党 |
| *laoqiang* (the sloping piece of wood ready for the protruding corner of roof) | 老戗 |
| Lian Po | 廉颇 |
| Liang Shanbo Temple | 梁山伯庙 |
| Liang Sicheng | 梁思成 |
| *lifang* | 里坊 |
| *lilong* | 里弄 |
| *lilü* | 里闾 |
| Lin Xiangru | 蔺相如 |
| *ling* (lattice) | 棂 |
| *ling* (spirit) | 灵 |
| *lingjiao yazi* (chevron-corbelled cornice on brick pagoda) | 菱角牙子 |
| Lingqiao Bridge | 灵桥 |
| Lingxing Gate | 棂星门 |
| Lingxing Star | 灵星，棂星 |
| Lin's Residence | 林宅 |
| longevity peach | 寿桃 |
| longevity rock | 寿石 |

| | |
|---|---|
| long-house-style building | 长屋建筑 |
| lounge bridge | 廊桥，廊屋式桥梁 |
| Lu Jiuyuan | 陆九渊 |
| *Luban Construction Treatise* | 《鲁班营造正式》 |
| Luocheng City (outer city) | 罗城 |
| M. A. N. Factory | （德国）孟阿恩 |
| Machinery Association of the Great Republic of China | 大中华民国机器公会 |
| Magnolia | 玉兰花 |
| main gate | 山门 |
| *maitou* (the corner part of building foundation) | 埋头 |
| major carpentry | 大木作 |
| Major Historical and Cultural Sites Protected at the National Level | 全国重点文物保护单位 |
| Mao County | 鄚县 |
| Maritime Silk Road | 海上丝绸之路 |
| Martino Martini | 卫济泰 |
| matchboard | 企口板 |
| Mawangdui Han Tomb | 马王堆汉墓 |
| Mazhu Station | 马渚站 |
| Mazu (Goddess of Sea) | 妈祖 |
| Measurement Map of Water System in Ningbo County | 《宁郡城河丈尺图志》 |
| meditation abode | 禅房 |
| Meiyuan rock | 梅园石 |
| memorial archway | 牌坊 |
| *Memories of the Construction and Salvage of Ningbo Lingqiao Bridge* | 《宁波灵桥兴建和抢救回忆志述》 |
| *Mencius* | 《孟子》 |

| | |
|---|---|
| *menzhen* stone (the stone placed on both sides of the gate to hold and fix the gate pivot and support the gate frame) | 门枕石 |
| merits and virtues arch | 功德坊 |
| *miangong shui* (sleeping bow-shaped water) | 眠弓水 |
| Miaogao Terrace | 妙高台 |
| "Miaogao Terrace" | 《妙高台》 |
| middle column | 中柱 |
| *mingceng* (*lit.* bright stories, or those that can be recognized from the outside) | 明层 |
| *Minglun Dadian* (*A Great Dictionary of Rites in the Ming Dynasty*) | 《明伦大典》 |
| Minglun Hall | 明伦堂 |
| Mingzhou | 明州 |
| Minister | 尚书 |
| Ministry of Justice | 刑部 |
| Ministry of Personnel | 吏部 |
| Ministry of Revenue | 户部 |
| Ministry of Rites | 礼部 |
| Ministry of Works | 工部 |
| Mogao Grottoes | 莫高窟 |
| *mojiao* | 抹角 |
| molding | 线脚 |
| mortise | 榫 |
| *mu* | 亩 |
| mud wall | 泥壁墙 |
| multi-eaved brick pagoda | 密檐式砖塔 |
| Nanchang Military Camp Design Committee | 南昌行营设计委员会 |
| National Highway G329 | 329 国道 |

| | |
|---|---|
| national intangible cultural heritage | 国家级非物质文化遗产 |
| Nationalist Government of the Republic of China | 国民政府 |
| *nenqiang* (the upturning piece of wood to make the protruding corner of roof) | 嫩戗 |
| *nianyu zhuang* (mainly blue and green, looking like the color of jade) | 碾玉装 |
| nightwatchman's drum | 更鼓 |
| Ningbo General Chamber of Commerce | 宁波总商会 |
| Ningbo Institute of Cultural Relics and Archaeology | 宁波市文物考古研究所 |
| Ningbo Commercial Group | 宁波帮 |
| Ningbo Postal District | 宁波邮界邮政局 |
| Ningbo Power Machinery Factory | 宁波动力机厂 |
| Ningbo Shipping Building | 宁波海运大楼 |
| Ningbonese Association in Shanghai | 宁波旅沪同乡会 |
| Ninghai Ancient Stage | 宁海古戏台 |
| "Ode to Lingguang Hall in the State of Lu" | 《鲁灵光殿赋》 |
| Officer of Metallurgy Affairs of Surveillance Commission | 按察使司金事 |
| "On Lingqiao Gate" | 《题灵桥门》 |
| "On Ten Major Relationships" | 《论十大关系》 |
| "On the New Water Clock" | 《新刻漏铭》 |
| "On the Steles of Chongde Yishu" | 《崇德义塾碑记》 |
| openwork carving | 透雕 |
| opera room | 戏房 |
| Osmanthus Hall | 桂花厅 |
| output rate | 出材率 |
| overhanging eave purlin | 挑檐檩 |
| overhanging gable roof | 悬山顶 |
| pagoda body | 塔身 |

| | |
|---|---|
| Pagoda of Tianning Temple | 天宁寺塔 |
| *pailou* | 牌楼 |
| Panchi Bridge | 泮池桥 |
| Panchi Pond | 泮池 |
| pavilion pagoda | 楼阁式塔 |
| Pengshan Pagoda | 彭山塔 |
| phoenix-peony stone column | 凤凰牡丹石柱 |
| *pingpan dou* (flat block) | 平盘斗 |
| pitched roof | 坡屋顶 |
| *pizuozuo* (splitting and building) | 劈作做 |
| plastering | 混水粉刷 |
| pointed arched doorway | 尖券形壶门 |
| pontoon bridge | 浮桥 |
| Port Affairs Section | 港务课 |
| pottery funerary object | 明器 |
| "Preface to Lanting Pavilion Collection" | 《兰亭集序》 |
| "Record of Rebuilding Lingqiao Bridge" | 《重修灵桥碑记》 |
| "Sanniang Teaches Her Son" | 《三娘教子》 |
| "Shu" (explanations to the notes) | 《疏》 |
| principal ridge | 正脊 |
| projected molding | 出挑线脚 |
| Provincial Administration Commissioner | 布政使 |
| Provincial Administration Commission | 布政使司 |
| Provincial Education Commissioner | 学使 |
| pyramidal roof | 攒尖顶 |
| *qi* (a six-*fen* gap or filler between two *cai*) | 契 |
| Qianfo Pavilion (thousand-Buddha pavilion) | 千佛阁 |
| *qianshi* (an official in charge of official business) | 佥事 |

| | |
|---|---|
| *qianzhuang* | 钱庄 |
| *qinglü yunzhuang* (green, blue and the blending of them as the transition) | 青绿晕装 |
| Qing'an Guild Hall | 庆安会馆 |
| Qin's Branch Ancestral Hall | 秦氏支祠 |
| Quan Zuwang | 全祖望 |
| *queti* (beak-shaped brace) | 雀替 |
| raised eaves | 升起，生起 |
| rear overhanging eave | 后出檐 |
| *Records on Examination of Craftsmanship* | 《考工记》 |
| red-lacquered gilded wood carving | 朱金木雕 |
| regular script | 正楷 |
| reinforced concrete structure | 钢筋混凝土结构 |
| relief carving | 浮雕 |
| reputable family in the east of Ningbo | 甬东名阀 |
| residential area and commercial port area for foreigners | 外国人居留地和商埠区 |
| ridge purlin | 脊檩 |
| Rinzai Zen | 临济宗 |
| Road of Ceramics on the Sea | 海上陶瓷之路 |
| roof truss | 屋架 |
| round carving | 圆雕 |
| round ridge roof | 卷棚顶 |
| *rulinlang* (a sixth-grade official) | 儒林郎 |
| *rupan* (*lit.* entering Panchi) | 入泮 |
| *ruxiang* (incoming of a minister) | 入相 |
| *ruyi* (good luck) | 如意 |
| Sage Confucius | 孔圣 |
| Shakyamuni | 释迦牟尼 |

| | |
|---|---|
| Sanbei Shipping Group | 三北航业集团 |
| Sanbei Shipport Company | 三北轮埠公司 |
| *sanheyuan* (a courtyard with buildings on three sides) | 三合院 |
| Sarira Hall | 舍利殿 |
| school map | 学宫图 |
| school-owned land | 学田 |
| screen wall | 照壁，影壁 |
| seal script | 篆书 |
| Secretary of a Bureau of a Ministry | 主事 |
| Semu (colored-eye) people | 色目人 |
| Senior Deputy Office of the former Zhejiang Customs | 浙海关高级帮办 |
| Shanghai Stock Exchange | 上海证券交易所 |
| *shanghui* (business association) | 商会 |
| *shi* (poems), *ci* (a form of classical Chinese poetry composed to certain tunes in fixed numbers of lines and words), *lian* (couplets) and *fu* (an intricate literary form combining elements of poetry and prose) | 诗，词，联，赋 |
| *shidian* (a memorial ceremony by offering alcohol and food sacrificel) | 释奠（丁祭） |
| *shidu xueshi* (a scholar in charge of the proofreading of *zouzhang*, the memorial to the emperor) | 侍读学士 |
| *shikumen* architecture | 石库门建筑 |
| Shipu Education History Exhibition Hall | 石浦教育史陈列馆 |
| *shique* (stone tower erected in front of palaces, temples and tombs to commemorate the owner's honors and achievements) | 石阙 |
| Shuibei Pavilion | 水北阁 |
| *shuiyue* lamp (the new-style gas lamp at the end of 19th century) | 水月灯 |
| Shuntian Prefecture (nowadays Beijing) | 顺天府 |

| | |
|---|---|
| *shuzhu* column | 束柱 |
| side door | 掖门 |
| Siemens Ltd. | 西门子（公司） |
| *siheyuan* (courtyard house with a fully enclosed courtyard) | 四合院 |
| sill boards | 槛板 |
| Siming Bank | 四明银行 |
| Siming School | 四明学派 |
| *Siming Topics* | 《四明谈助》 |
| *single building* | 单体（建筑） |
| single-arch stone bridge | 单孔石拱桥 |
| single-side columned veranda | 单面列柱外廊 |
| Siqian Bridge | 寺前桥 |
| sitting room | 中堂 |
| *Six Annuals of Siming in the Song and Yuan Dynasties* | 《四明宋元六志》 |
| Six Kilns | 六大窑系 |
| slash-and-burn cultivation | 刀耕火种 |
| solid wall | 实砌墙 |
| Son of Heaven | 天子 |
| Soto Zen | 曹洞宗 |
| southern operas | 南戏 |
| Southerner | 南人 |
| spinning wheel | 纺轮 |
| Spring and Autumn period | 春秋时期 |
| State of Wu | 吴国 |
| stele inscription | 碑记 |
| stepped gable | 马头墙 |
| Steward-bulwark of State | 宰辅 |

| | |
|---|---|
| stilt architecture | 干栏式建筑 |
| stone adze | 石锛 |
| stone-structured five-arch ring bridge | 石结构五洞环桥 |
| street stage | 街心戏亭 |
| Sub-prefectural Magistrate | 知州 |
| Sumeru base | 须弥座 |
| Sun's Siben Hall | 孙氏思本堂 |
| Sunjiajing Ancestral Hall | 孙家境祠堂 |
| *suoshi* (stone lintel beams) | 锁石，横楣梁 |
| *Supplementary Annals of Fenghua County* | 《奉化县补义志》 |
| Svaha Liu | 刘萨诃 |
| *Taifu* (Grand Mentor in charge of civil affairs) | 太傅 |
| *taiming* (the surrounding edge surface and outside walls of the building foundation) | 台明 |
| Taiping Army (army of Taiping Tianguo, or the Taiping Heavenly Kingdom) | 太平天国军 |
| Taiping Rebellion | 太平天国运动 |
| *tangqian* (*lit.* in front of the main hall) | 堂前 |
| *tanhua* (a scholar who ranked third in the highest imperial examination) | 探花 |
| Taoism | 道教 |
| Taoist temple | 道观 |
| tapering | 收分，收杀 |
| Tax Department | 税务司 |
| technical director | 技正 |
| Temple of Guan Yu | 关帝庙 |
| Temple of Qimu Jiangjun | 七牧将军庙 |
| temple school | 庙学 |
| Ten Views of Qianhu Lake | 钱湖十景 |

| | |
|---|---|
| tenon and mortise joinery | 榫卯（结构） |
| terrazzo | 水泥磨石子 |
| the 14 Creditable Officials of Lingyan Pavilion | 凌烟阁十四功臣 |
| the ancient county *yamen* | 古县衙 |
| *The Annals of Cixi County* | 《慈溪县志》 |
| *The Annals of Ningbo* | 《宁波市志》 |
| *The Annals of Yin County* | 《鄞县志》 |
| the art of carved painting | 雕花髹漆 |
| the Bank of China | 中国银行 |
| the Banking Guild Hall | 钱业会馆 |
| the basin-shaped column base | 覆盆状柱础 |
| the Beiyang Army | 北洋军阀 |
| *The Book of Changes* | 《周易》 |
| the British Consulate | 英国领事馆 |
| the Chinese People's Political Consultative Conference | 全国政协 |
| *The Chinese Speaker* | 《官话》 |
| *The Collection of Baiyun* | 《白云集》 |
| the first and fifteenth day of each month | 朔望 |
| the Four Greatest Calligraphers in the Early Tang Dynasty | 唐初四大家 |
| *The Garden of Words* | 《字苑》 |
| *The Geological Survey of China* | 《中国地质约测图》 |
| the Grace and Glory of the Emperor | 恩荣 |
| the Grand Secretariat | 内阁 |
| the highest imperial examination | 殿试 |
| *The History of the Song Dynasty* | 《宋史》 |
| the Imperial Bank of China | 中国通商银行 |
| the middle layer of the Sumeru base | 束腰 |

| | |
|---|---|
| the Moon Lake | 月湖 |
| the National Chamber of Commerce and Industry | 全国工商会 |
| the Neolithic Age | 新石器时代 |
| "The Night View from Lingqiao Gate" | 《登灵桥门晚望》 |
| the North Riverbank | 江北岸 |
| the Northern Expedition | 北伐战争 |
| the Old Bund | 老外滩 |
| the period of modern (Chinese) history (1840–1949) | （中国）近代史 |
| *The Protection Planning of Shipu Historical and Cultural Reserve* | 《石浦历史文化保护区保护规划》 |
| The Queen of Heaven (Mazu) Temple | 天后宫 |
| the period of the Republic of China | 民国时期 |
| *The Rites of Zhou* | 《周礼》 |
| *The Romance of the Three Kingdoms* | 《三国演义》 |
| the seal of the literary chamber | 文房之印 |
| the Shanghai General Chamber of Commerce | 上海总商会 |
| *The Story of Pipa* | 《琵琶记》 |
| the Swire Group | 太古洋行 |
| the temple of Taoism gods | 境主庙 |
| the theory of continental drift | 大陆漂移学说 |
| the Three-river Junction (the three rivers are Fenghua River, Yaojiang River and Yongjiang River) | 三江口 |
| the Tower of Wenfeng (*lit.* Peak of Literature, which is believed to bring good luck to those taking literary examinations) | 文峰塔 |
| the upper intermediate purlin | 上金檩 |
| Thomas Young | 汤姆士·扬 |
| Three Carvings (wood carving, stone carving and brick carving) | 三雕 |

| | |
|---|---|
| Three Gods "*Fu*, *Lu* and *Shou*" (Gods of Happiness, Fortune, and Longevity) | "福、禄、寿"三星 |
| three promotions in a row | 连升三级 |
| Three Steles of Operas | 戏剧三通碑 |
| Tianfeng Pagoda | 天封塔 |
| *tianjing* courtyard (*lit.* sky well) | 天井 |
| Tiantong Temple | 天童寺 |
| Tianyi Pavilion | 天一阁 |
| tie-beam | 枋 |
| tile wall | 瓦爿墙 |
| timber-frame architecture | 木构建筑 |
| *tixue qianshi* (an official in charge of education) | 提学佥事 |
| Todaiji Temple | 东大寺 |
| town house | 联排式住宅 |
| Treaty of Nanking | 《南京条约》 |
| treaty port | 通商口岸 |
| tribute student (one of the top performers designated under the Directorate of Education for further study and civil service) | 贡生 |
| triple caisson ceilings | 三连贯藻井 |
| triple-side columned veranda | 三面列柱外廊 |
| trough plates with a circular-arc groove on both sides | 双桦槽板 |
| truss | 桁 |
| T-shaped *gong* | 丁头栱 |
| Tuoshan Weir | 它山堰 |
| Twelve Philosophers | 十二哲 |
| two-step cross beam | 双步梁 |
| upper architrave | 阑额 |
| Vajras (Buddha's guardian warriors) | 金刚 |

| | |
|---|---|
| veranda lobby | 外廊式门厅 |
| veranda-style architecture | 外廊式建筑 |
| vertical ridge | 垂脊 |
| Vice Censor-in-chief | 副都御史 |
| Vice Director (of a ministry) | 员外郎 |
| Wang Yangming | 王阳明 |
| *wanjie* (the completion of duty with moral integrity) | 完节 |
| water beam | 水梁 |
| water clock (an ancient timer) | 刻漏；漏壶 |
| Wen Temple (temple in honor of a philosopher, Confucius) | 文庙 |
| Weng Wenhao's Former Residence | 翁文灏故居 |
| *wenshou* (zoomorphic ornaments) | 吻兽 |
| Western town house style | 西方联排式 |
| Western-style garden house | 花园洋房 |
| Western-style Roman column | 西式罗马柱 |
| width | 面阔 |
| wing building | 翼楼 |
| wing room | 厢房 |
| wing wall | 八字墙 |
| winged arrowhead with long cutting edges | 双翼长锋箭镞 |
| winged arrowhead with short cutting edges | 双翼短锋箭镞 |
| Woji Hut | 蜗寄庐 |
| *wokou* (Japanese pirates) | 倭寇 |
| wood-cored lacquer bowl | 木胎漆碗 |
| wooden boards with dowel pins | 销钉木板 |
| wooden caisson | 木藻井 |
| wooden corbels | 牛腿 |

| | |
|---|---|
| wooden gate stops | 门档 |
| wooden spade-like farm tool | 木耜 |
| wooden stilt | 桩木 |
| wood-like brick carvings | 仿木砖雕 |
| worshiping the ancestors | 崇宗祀祖 |
| Wu Temple (a temple in memory of a military saint, Guan Yu) | 武庙 |
| *wucai bianzhuang* (blue, green, red and other colors all over the components in the architecture) | 五彩遍装 |
| Wufeng Building (*lit.* five-phoenix building) | 五凤楼 |
| Wuji Bridge | 戊己桥 |
| *Xiangshan Journal* | 《象山日志》 |
| Xiangxian Temple | 乡贤祠 |
| Xiansheng Temple stage | 仙圣庙戏台 |
| Xiantong Pagoda | 咸通塔 |
| Xiaoxi rock | 小溪石 |
| Xie Hengchang | 谢恒昌 |
| Xie's Ancestral Hall of the First Ancestor | 谢氏始祖祠堂 |
| *xiucai* (those who passed the country-level imperial examination) | 秀才 |
| Xi'ao Stone Arch Bridges | 西岙石拱桥 |
| *xinxue* (School of Mind) | 心学 |
| *Yamen* (the headquarters or office of the head of an agency) | 衙门 |
| Yan Xinhou | 严信厚 |
| Yan Zijun | 严子均 |
| *yanbu* | 檐步 |
| *yangma* (the major beams) | 阳马 |
| Yangtze River Delta | 江南水乡 |

| | |
|---|---|
| Yanyu Building | 烟屿楼 |
| Yao Mo's Former Residence | 姚镆故居 |
| Yaohang Street | 药行街 |
| Yaojiang culture | 姚江文化 |
| Yellow River Basin | 黄河流域 |
| *yimen* gate | |
| Yin (County) | 鄞县 |
| Yin County Council | 鄞县县议会 |
| Yin County Literature Committee | 鄞县文献委员会 |
| *yingqing* porcelain (*lit.* shadow celadon porcelain) | 影青瓷 |
| Yinzhou District | 鄞州区 |
| *yishu* (free private school) | 义塾 |
| *yongdao* (path) | 甬道 |
| Yongfeng Warehouse | 永丰库 |
| Yongju opera | 甬剧 |
| Youth Palace | 青少年宫 |
| Yu Qiaqing's Former Residence | 虞洽卿故居 |
| Yuanfengrun Draft Bank | 源丰润票号 |
| Yuantong Customs Official Bank | 源通海关官银号 |
| *yuantuo* (a kind of turtle in the Chinese myth) | 鼋鼍 |
| *yuanyang jiaojing gong* (the bracket arm that looks like two mandarin ducks crossing their necks in love) | 鸳鸯交颈栱 |
| *yueliang* beam | 月梁 |
| Yuling Tomb | 裕陵 |
| *yunban* (cloud-shaped iron board used for announcement of public notices) | 云板 |
| "Yundong Zhuzhi Ci" | 《鄞东竹枝词》 |
| *Zhangyuanshi* (the official in charge of the academy affairs in Hanlin Academy) | 掌院事 |

| | |
|---|---|
| Zhejiang Customs | 浙海关 |
| Zhejiang Industrial Bank | 浙江兴业银行 |
| Zhengren Lecture | 证人讲会 |
| *zhengshi* (soldier on expedition) | 征士 |
| Zhengyi Lougong Lecture House | 正议楼公讲舍 |
| Zheng's Chongde Hall | 郑氏崇德堂 |
| Zhenhai Estuary | 镇海口 |
| Zhenhai Tower | 镇海楼 |
| *zhiliangzhi* (extension of innate knowledge of the good) | 致良知 |
| "Zhonghao Yihui" Stone Pavilion | "钟郝遗徽" 石亭 |
| *Zhongshu* (secretary, in charge of drafting, recording, translating and hand-copying official documents) | 中书 |
| *Zhongshu Lang* (inner secretarial court gentleman) | 中书郎 |
| *Zhongshu Sheren* (imperial secretary, mainly in charge of drafting imperial mandates at the Palace Secretariat) | 中书舍人 |
| Zhongtian Pavilion | 中天阁 |
| Zhou Jinbiao | 周晋镳 |
| *zhouxue* (prefecture school) | 州学 |
| Zhu Shunshui | 朱舜水 |
| *zhuangyuan* (Number One Scholar, a title conferred on the one who came first in the highest imperial examination) | 状元 |
| *zhusheng* (alias *xiucai*, referring to those who passed the county-level imperial examination) | 诸生 |
| Zicheng City (inner city) | 子城 |
| *zoumalou* building | 走马楼 |
| Zunjing Pavilion | 尊经阁 |

Translator's Note

# 译后记

　　《奇构巧筑：宁波建筑文化》（*Architectural Wonders in Ningbo Land*）作为宁波文化丛书之一，是黄定福先生十余年前所著关于宁波建筑历史和发展的一部书。对于想了解宁波建筑文化甚至中国古代建筑的西方读者来说，这是一本难得的入门读物。这本书不仅是一部宁波建筑文化的发展史，也是中国建筑文化发展史的一个缩影。

　　建筑既体现民族性，又兼具时代感，正如梁思成先生在《图像中国建筑史》一书中所说："建筑不仅是一些砖头和木料而已，它是一门艺术，是民族和时代的表征，是一种文化遗产。"建筑见证历史变迁，建筑见证文化演变。让西方了解一个城市的历史和发展，不妨让他们来了解这个城市的建筑。于是，将介绍城市建筑发展史的文字翻译成英文就显得很有必要了。作为建筑专业的门外汉，这大概是译者第一次真正靠近中国建筑吧，尤其是传统古建筑。门外汉译专业书，难于古人走蜀道。主要在于"两难"：一是建筑术语难，二是诗文楹联难。首先，原著中的建筑类专业术语，知之不易，译之更难。所谓知之不易，是说译者对于其中生僻的建筑术语不能理解，在不理解的情况下，翻译根本无从下手，只能请教专业人士，或者翻阅建筑书籍，或者通过网络查阅大量的相关资料。而即便如此，常常一个术语花上好些天工夫也未必能找到满意的解答。所谓译之更难，是说原著中的建筑术语和诗文楹联要找到对应的英文表达，更远非一件易事。

　　所以，磕磕绊绊完成译稿的两年半漫长时间里，首先要特别感谢梁思成先生。他的两部著作，《图像中国建筑史》（*A Pictorial History of Chinese Architecture*）和《为什么研究中国建筑》（*Chinese Architecture: Art and Artifacts*）为译者提供了深入浅出、隔行易懂的专业知识以及许多生僻难懂的专业词汇对应的英文表达，帮助译者在很大程度上解决了一部分基本专业词汇的理解和翻译。由于书中的古建筑术语所指大多是中国传统建筑所特有的构件，属于汉语独有的文化负载词，因此它们缺乏对等的英语词汇。其中大部分术语梁先生采用音译法，但他使用的是那个时期比较普及的威妥玛拼音，考虑到这套拼音系统已经被现代汉语拼音系统所取代，译者在选择音译策略时放弃了梁先生的威妥玛拼音而换成了现代汉语拼音斜体的方式，有的生僻术语则在拼音后面加括号注释。

我的同事，宁波大学科学技术学院建筑工程学院的赵圣洁老师在专业术语的翻译过程中提供了重要的帮助。她主攻外国建筑，中国传统建筑并非她专业所长，但她每次都一定会倾力相助，帮我查找大量资料，或帮我咨询中国建筑的专家。

诗文楹联部分，则要感谢人文学院特聘院长周志锋教授。众所周知，中国古代建筑与文学艺术之间存在着密不可分的共生关系，诗词、楹联、匾额、碑文等这些叠加在古建筑上的文字，与建筑本身相得益彰，增添了建筑的意境美和想象的空间，形成中国古建筑一道独特的风景。《奇构巧筑》也不例外，涉及了大量的诗词、对联和古文。为了实现翻译最重要的语义对等，译者向周志锋教授一一求证请教。每次把疑难问题发给他，只要他在线（长期伏案写作的他很少不在线），总是能够第一时间给予反馈；若有不确定之处，他便告诉我收到了，但要稍后回复；若是没能及时回复，他必定会诚恳地表示歉意："抱歉，今天……迟复为歉。"对于我的疑惑，周老师不仅逐字逐句详细分析解释诗词古文的意义，有时候发现原文与实际有出入的地方还会去查阅大量的资料，避免以讹传讹，哪怕只是一个错别字，或者仅仅一个印刷错误。周老师的解答连标点符号都不会放过，还会附带把追根溯源的出处一并发来，或网址，或网络截图，或词典拍照。周老师作为一名学者的严谨治学求真精神让我肃然起敬。

封面的河姆渡遗址干栏式建筑图片，是由北京电影学院动画学院本科毕业、旧金山艺术学院硕士毕业的周鹏先生于百忙之中挤出时间手绘制作，并免费提供给本书使用，我感激不尽，在此向他致以特别的谢意！

本书初稿译出后，有多年翻译经验的董铁柱老师也给予了很多具体的专业修改意见，也让我在翻译技巧和水平上有所精进。责任编辑黄静芬老师和其他参与审校工作的老师也为本书的出版不厌其烦地做了许多细致的工作。在此向这些专业而敬业的老师表达我的感激。我可爱的学生，宁波大学科学技术学院人文学院 2021 级英语专业空乘 1 班的商琪同学在术语表这一部分也帮我做了大量的整理工作，非常高效，对待细节很认真。诚表谢意！还要感谢宁波大学科学技术学院人文学院 2020 级翻译专业本科生钱丽霞、2021 级翻译专业本科生林仪，以及宁波大学外国语学院 2022 级翻译专业硕士研究生杨澜、李慧琳、吕梦怡，她们为本书做了大量的资料查询、文字校对、术语整理等工作。

感谢家人对我的支持，为我付出的一切。我的父母，他们洗衣煮饭除尘整理，帮我免去了料理家务所需的大量时间。我的先生，虽常驻外地工作，但非常理解我工作的辛劳，常常抽空回家帮忙料理孩子的学业和生活，让我专注于教学和手头的翻译工作。

翻译初稿的完成是在 2023 年的 1 月 18 日。那几日恰好感染上新冠，自我隔离在卧室，也隔绝了门外的各种干扰。慢吞吞的翻译工作也因此顺利接近尾声，而 17 号夜晚正好表现出失眠的症状，从来一秒入睡的我彻夜难眠，数番辗转后决定爬起来继续翻译。终于在第二天清晨 5 点的曙光中，初稿出炉了。距离接手这一翻译任务恰好一年左右。

最后要感谢人文学院的贺安芳院长和项霞副院长，让我有幸承担起这份重任，作为隔行勉力将本书译出，为普及宁波乃至中国建筑史的知识尽一份微薄的力量。本书书名的英译 Architectural Wonders in Ningbo Land 便是项霞老师提供的建议。

黄定福先生的这部作品本身便是一座璀璨的文化宝库，能够参与其翻译工作，我深感荣幸。此次翻译对译者来说是一次宝贵的学习经历，不仅加深了译者对建筑，尤其是中国古建筑的了解，也让译者在翻译技巧上有所精进。在翻译过程中，译者虽力求忠实于原文，但毕竟水平有限，也深知语言与文化之间的鸿沟难以完全跨越，难免存在诸多不尽如人意之处，恳请方家和读者不吝指正。

毛智慧

2024 年 7 月 25 日于宁波

2024 年 10 月 28 日补正